It's the Liver Stupid *Ancient Healing Arts Combined with modern day science to deliver health and youth..5th Edition* *12/29/17*

Table of Contents:

The Author at the helm of a Jeanneau on a leisurely downwind leg in British Virgin Islands . . . His passion

Foreword

The 4th and 5th editions of this Book compiled and written in 2017 are significant updates, but the first edition was written over a six-year period starting in about 2000. It began with the Yahoo online forums below and both are currently still functioning as of 2/17:

Coconut Oil http: /5groups.yahoo.com/neo/groups/coconut_oil_open_forum/info
I first learned of High Level MSM in 1998 and this group, started in 2006. The High Level MSM discussion online started maybe three years prior in the Beck group below.

Beck Blood Electrofication was started in 2001
1http://groups.yahoo.com/neo/groups/Beck-blood-electrification/info

These discussions inspired this manuscript and the first comments stating with Chapter 52 go back to 2007 when I just started seriously reporting my findings and compiling the information. The original discussions can be still be researched in the two above mentioned Group Archives. I have cleaned up some of them, just to avoid confusion, but they are generally left in tact with spelling and punctuation errors and little editing. Those familiar with the forums know that members often write somewhat in shorthand, but the meaning generally comes through.. Where they don't, I have attempted to clean them up without losing their originality, but keeping their thought and content.

While the body of this book is only about 140 pages, if you follow the links, and I recommend it highly, you could literally spend a week or more reading and watching the referenced videos. Some of the people who post are healthcare professionals, but most are people searching for ways to solve their own health problems.

To master this AHC program, one must watch the videos and read the online material, which serve as the bibliography. Some of these can just be scanned, but other parts, like the Seneff interview are essential to getting the whole picture. I have done my best to sort out the most pertinent parts, but the pieces that I drew from are all still in tact in these references, if you are interested.

As the information here is constantly changing as are web addresses, editing it often to keep it current is essential. Some of the web addresses change and that is the problem with a live book. However, the reader is encouraged to look up the references as they are often key to the story here and a part of the education that leads to wellness and longevity.

I am especially indebted to: My friend and past client, Dr. Deborah Hobbs, local chiropractor, who taught me some of the healing ropes in the late '90's. Duncan Crowe, a Canadian Nutritional Expert and Ozone Therapist, who now is retired in the Canadian Rockies, but once sold several lines of food supplements that he has creatively researched. His Whey Protein protocol is basically a result of his research and I give him full credit for sorting it out. Also, I thank my past business partner, Dr. Kevin Freeman, M.D., a specialist in Endocrinology and a Pharmaceutical Consultant. He knew the other side of this health story about as well as anyone and gave me a great deal of help and background early on.

Finally, thanks to my friend and co-worker on several health ventures, especially water technology, Dr. Richard Price.

Dedication

This 4th &5th edition is dedicated to my loving brother Steve who edited the 1st edition and died in 2016 of lung cancer in a crazy struggle that could have easily been avoided.

1. Introduction

This is a multimedia book on the Anti-Aging and Health Program (AHP). Originally, it focused just on MSM (methylsulfonylmethane), but with my research, a light went off. On reading some very deep scientific papers, I realized that the real key to health rests with the liver, and that joint health is one small fraction of a much bigger issue. The 4th Edition brought out the methylation component where the liver is further the point of focus. With this, I combined some of the other powerful foods and food supplements that I have come across, mainly for reader feedback. They all affect the Liver, generally, but not all effect it directly. Now I realize that the unconscious mind helps us master the master organ.

Please read this note below, focus on Mona's problems and the resolutions:

You don't know me, but I bought the Horse MSM a year ago per the protocol.

I've had 4 spinal fusions and a hip clean-out. I have Crohn's and PSC. PSC is an auto immune bile duct disease. The only TREATMENT is liver transplant. I've already gone from stage 4 to 0-1. I WAS taking 14 pharmaceutical medications, and now I take ZERO.

My doctors HATE it, but they can't argue with success. We are nearly over diagnostics (I just use them for testing to help myself). I can then say: OK, are we done. YES, then I will discuss what I have been really doing all along with them.

Thank you for this… you have NO IDEA! Mona, age 54, Apple valley, CA

Now please watch Dr. Tent's Video that Mona sent me as a result of our discussions:
http://tv.greenmedinfo.com/dr-tent-the-exploding-autoimmune-epidemic/?utm_source=www.GreenMedInfo.com&utm_campaign=85ce75b581-Greenmedinfo&utm_medium=email

So HL MSM is working just as these combined theories predicted, but no one had any idea that it could affect Crohn's and PSC. However, at the time of this letter, I really knew nothing about the methyl (**MSM**) part of the puzzle nor how it affected autoimmunity and cancer. When you do, the cures fall into place. The methylation pathways are where it all begins (5th edition). None of this is even alluded to by the MSM experts who have sprung up on the web since the first edition of this book.

At the end, I include Yahoo Coconut Oil and Beck-blood-electrolysis Forum "posts" that reflect the history of my understanding of all of this. Following them are more recent posts that reflect on and contribute to a deeper understanding as I answer questions that Group members addressed to me.

Again, to master this program, one must watch the videos and read online material, which serve as the bibliography. Some of these can just be scanned, but other parts, like the Seneff interview are essential to getting the whole picture. I have done my best to sort out the most pertinent parts, but the pieces that I drew from are all still in tact in the group archives, if you are interested.

This book first started when the architectural market fell flat in '07. I began by compiling the feedback and comments from members of our Groups. It has grown from a labor of love to, more recently, a passion, as interest has grown and people world over have reported cures like Mona's... from Arthritis where it started to Cancer and Heart Disease.

How this book began... and my inspiration

I am a professional architect, not a healthcare professional, and I make no such claims. My interest in health is with regards to staying well, which means staying young, not in chronological years, but in response to aging factors that we consider normal.

Furthermore, this interest in health was first inspired by my father. Dad was an Environmental Engineer (called Sanitary Engineer in those days) with degrees in Chemical and Civil Engineering, but his passion was always in medicine. He lived before the internet was conceived, but he was a whirlwind of knowledge and would read as many as six or seven newspapers per day in an effort to sort out the truth from prevailing fiction. This was difficult to do then, really impossible to do with the controlled media today, but now more possible with the world wide web to help us if we can keep a step ahead of the controlling search engines, viruses, and media hype..

Dad knew more about physiology and anatomy than most doctors, but unfortunately, he was duped by the mainstream and actually bought into the Lipid/ Dr. Keys' story that I discuss herein. It was not pretty. He literally blew up at age 67 of a massive heart attack and stroke. Its too bad that I cannot bring him back into this discussion.

But if he were here today, I am sure that he would be all over the things that I reveal herein and have a keen interest in with the new developments that are coming out almost daily in nutrition, antiaging, and health. None of it stems from government, by the way.

While my story starts with a gift based in horse farming and veterinary medicine, it quickly moves to, especially, the Sciences of Anti-Aging and Body Electronics, then, beyond, to the metaphysical realms and finally to Pure Spirituality. It is a wild ride, but it all fits together and tells a compelling story that can lead just about anyone to optimum health and beyond, that is, if they pay attention and follow it carefully.

2. Background

I am my own patient and was born with juvenile arthritis in my fingers and hip (my friend, Dr. Freeman's diagnosis based on my description). By 1976, I could not throw a ball more than thirty feet and, as a result of a ski accident, my knee was badly injured that year. My hanging over drawing boards for hours on end, stressed my neck terribly. I had nearly unbearable joint problems and pain everywhere. By the year 1983, I was a physical mess and wearing a neck brace, most days, to work. This I considered a God-send. Interestingly, my brace was often borrowed by my office cohorts.

My physician at Harvard Health Center in Boston described my neck as a work-related problem common to architects, but I now know that joint problems are much more deeply rooted in most all cases. More accurately, after putting the very deep scientific relationships and studies together in writing this book, I have come to understand that such conditions could easily be the result of Liver Stress. Furthermore, I discovered that this generally begins to, strongly, manifest at around age 40 for western High Carb Diet/ High Sugar civilizations where I squarely fell at the time

Some of the above problems subsided somewhat over time as the health care professionals advised that they would, but others, like sciatica, cropped up in their place. By 1992, at age 50, they had virtually all reappeared. In addition to these, I had circulatory and other health issues that seemed unimportant, by comparison, but now I see as life threatening warnings that I, like most people, ignore. I was destined to early death... following my father.

Also, like most people, my concerns were mainly with the joint pain issues. The most immediate of these around this time was sciatica. Sciatica pain, if you have never encountered it, virtually takes away your breath when it hits . . . it can make you cry.

One day then, I was throwing the ball for my two-year-old golden retriever, Axel, and the sciatica hit me. Crying out in pain and tears, I realized that no one could help. So I crawled into the house. Axel tried to comfort me, but he couldn't.

Sciatica pain, I learned, would typically last about a week, but it would slowly subside on its own and, generally, disappear only to reappear again in about six months and with regularity. I did seem to get faster recovery with help from my Chiropractor friend, Dr. Hobbs. However, this was indeed something that one does not forget. Furthermore, no matter what, one can expect it to come back within a year, in any event, once you are prone to it. Until the direct attack, it comes on just like the hip pain that I was born with. It is a low level ache in the leg to hip joint... As explained more carefully by Dr. Hobbs, it's an inflammation of the sciatic nerve that leads to the leg that finally comes full-force with a crushing attack. In any event, no matter how carefully it is explained, I can guarantee that you don't want it explained to you.

Sciatica was bad, but it was not the worst pain that I have ever encountered. At age 14, I had my first migraine. I later learned from my son's doctor that this type is generally called a "Classic Migraine." Others refer to it as an "Optical Migraine."

I still got mild migraine attacks occasionally until about 2008, but in those early days, they were near-death experiences. I attribute them now to mercury fillings in my teeth, but there is no way to prove where they originated. It was not until after spending four years in the Air Force that I had even a clue as what they were. They are not really headaches, but once the attack subsides, you are blessed with just a headache. Military doctors would stare at you blankly, pass you some green "G.I. gin," and send you to the barracks to throw up and endure the pain, which would normally just last one day.

My knee, a severely torn ACL, got better. However, by age 50, when I played tennis, it started snapping out of the joint. I would land on the court in a pile until I could recover myself. I saw several MDs over this knee condition. The last one, Dr. James Marvel, a highly regarded local sports specialist, told me that he had the same problem, and that knee replacement was, at most, two years away. Fortunately, I had just watched a documentary on PBS describing what appeared to be carpentry... but passed as surgery. This did not and does not now fit my vision of medicine and healing.

Furthermore, as an architect plus accomplished structural designer, yacht & furniture builder, carpenter, and auto mechanic, I have a very good understanding of stresses. There is no real source of lubrication in an artificial joint. It is a separate foreign entity and one that I can see is bound to eventually wear out and fail.

My vision of an artificial joint failing was always this: You are walking or running and the hip or knee joint fails. You fall down in a pile with this piece of stainless steel protruding from your body with a cleanly separated joint, looking like someone hit it with the blunt end of a meat cleaver, as you writhe in pain... are you getting this? It's not going to be a pretty picture when this artificial entity falls apart as it must unless you die first as medicine depends that you will.

Fortunately, I came across a bulk MSM salesman on dial-up in about 1998. A down-to-earth and very intelligent farmer with a crazy yellow website explained that he sold MSM for horses, not people, and by the rail car load. He told me that his family of over thirty people had no joint problems: no knee problems, no sciatica, no neck pain, no arthritis of any kind, and, for the most part, seemed more healthy than other people in their community in every way. I am probably not giving this man enough credit, but I was immediately intrigued.... and I am grateful that I was paying attention that day.

He then told me that the key was MSM. I explained to him that I had been taking MSM for years... and that he was right, it worked, but not as well as he was describing. He then revealed what I am describing herein as the "HL MSM Protocol." From here, he then agreed to bag up ten pounds of his product and send it to me for his cost. I have never looked back. Within two months, virtually every joint problem that I had been enduring was gone. The sciatica never came back. Still, some of the rheumatoid arthritis took a bit longer. I was not particularly aware of it at the time, but, with the, other conditions, like poor circulation and migraine headaches, slowly began to disappear also. They were all somewhat related. Who would have thought? I now knew that this was no coincidence. I have gone from a sick ragamuffin to being in better health than I was at age 27 when I could not drive a car with my hip flair-up. I had virtually aged backwards. Today, I do not even catch colds, nor do I ever get flu or any viruses. I am simply well as nature intends your body to be and it keeps getting better as years pass.

Since meeting this man online, I have always been very interested in what MSM really does, its characteristic, its derivation, etc. But, from that moment on, I have never doubted its contribution to my overall heath and to my health in general. I have been spreading the word about this high level MSM in the two above-mentioned discussion groups, but even before that, in chat rooms, on dial-up. In doing that, I have also learned from others, especially Duncan Crowe whom I mention, herein, often. Duncan introduced me to many of the co-factors that I reintroduce herein but with added knowledge as I study them in depth.

It was Duncan who first made me aware of the Coconut Oil Group and invited me to join. While Duncan is not an M.D., he is (was) a unique and gifted Health Care Professional. To say the least, he is very talented at what he does. His special interest is in Anti-Aging and its implications. He has put a great deal of effort there in his personal investigations. He is also an Ozone Therapist and sells a number of creative products that he had researched well on his website.

At times, Duncan's scientific bent never allowed him to accept word of mouth as proof and he was always looking to PubMed as a source of information. His once extensive website no longer exists, sorry to say, and he has retired to the Canadian Rockies.

Duncan had me looking into DMSO from his experience with his Father. However, it was not until the Dr. Stephanie Seneff in Dr. Joseph Mercola interviews, a few years later, that I began to understand, fully, the significance of the Organic Sulfur component of MSM. Dr. Seneff, a brilliant scientist/ engineer, without her knowing it, has unlocked one of the several keys to this amazing organic chemical/ food that is oxidized from DMSO. DMSO is (was) derived from wood products. MSM and DMSO can be used in different ways, in joint treatment and general health. However, DMSO is much more difficult to use safely as it pulls other chemicals in with it. With the 4th edition, I understood them both at the microcellular level, where the real work is done and explain their real workings at this depth herein, but sulfur is still a key factor in this story.

Also, while the HL MSM Protocol is generally safe as I explain later. Please read the Quick Start section, always... Take your time and listen to the responses from your body as you go, on all levels. Still, MSM is most assuredly the most safe of all food supplements even if you avoid my warnings, Take it incorrectly, hate the taste, and, still, you will likely get better, no matter what your health issue happens to be. If there is a magic bullet in supplements, this oxygen transport pair is the representative.

3. AHP

Welcome to the **Anti-Aging & Health Program**. I hope that you find this program description and component parts thought provoking and accept them with the same level of enthusiasm that I have. It's a multimedia adventure that should take you a lot longer to fully complete than you need to "just to get the show on the road."

To start, I suggest that you read the basics of the **HL MSM Protocol** below, then skim through everything. Follow that by ordering any of the other foods/ parts that jump out at you. Organic C (or Liposomic C today) is a valuable cofactor in the process, but there are many more. When done, come back to this point and start your comprehensive read. Please take your time to research each live link as I have... by the time MSM (and whatever else) arrives, you should have a pretty good idea of how to use this AHP properly. I am totally amazed as to how this program fits together so seamlessly. The comprehensive research that fits into every level of these components is amazing, but the brilliant researchers who did the groundwork obviously do not talk to each other. I do this for you, so follow me here. They somehow missed the connections between their work and that of others. These were generally obvious to me, once I got through them, but I found the connections inspiring. I only hope that I have pointed them out clearly to you. Still, you can reach me through these groups if you have questions or drop me an email.

Enjoy!

The Quick Start

The basics of the **HL MSM Protocol** are very simple and this is always what people ask for first:

- **Go to Amazon, Search MSM for horses and buy 10lbs of it (for about $45 with S&H)**
- **Start out at about a heaping teaspoonful (amounts are never critical).**
- **Take a scoop with water, swish it around in your mouth for at least 3 min (that's a long time). There should be no taste at this point and for the very good reason that I point out later. At this point, just be conscious that this stuff is affecting you at the microcellular/ mitochondrial level in transformative ways that you could not imagine.**
- **Swallow it.**
- **If you have no negative reaction after a week at this level, double it... Work up to the point where joint pain (or whatever your issues) never occurs.**

Now go to step 1. before you read any further. Order the MSM @:
http://www.amazon.com/AniMed-PureMSM-10lb/dp/B000HHQ5BE/ref=sr_1_2?ie=UTF8&qid=1358611984&sr=8-2&keywords=msm+for+horses

While you are at it, buy five pounds of Now Brand Whey (for the simple folks who are not so concerned with the radioactivity from the Japanese reactors like myself. If that concerns you, search the web for MSM from New Zealand that he is currently recommending:
http://www.amazon.com/Foods-Whey-Protein-Isolate-Packaging/dp/B0015AQL1Q/ref=sr_1_cc_1?s=aps&ie=UTF8&qid=1359589874&sr=1-1-catcorr&keywords=now+whey
(this is just one of hundreds of choices)... and get started. Your future in good health depends on it. If you are not so frugal, go to Swanson Vitamins and buy Goat Whey.

Your Body Knows

OK, you have ordered your MSM (and Whey)... Now what? First, we must understand that this AHP (Anti-Aging & Health Program) when fully understood and implemented will be "your" individual program and not just a diet, a supplement, or any single protocol. Note that neither here nor in my later books do I recommend single solutions. One size never fits all. We are individuals. (5th editions).

By the time you have waded through the comprehensive videos, etc., you will have molded it to fit your body. My hope is that you will have also begun to access your intuitive inner guidance in gathering this. Your spiritual as well as your body's feedback can then work together and are the key to this totally unified program.

In this process, from the physical side, you heed the warning signs of pain at the one extreme. At the other, you begin to develop a knowingness that will guide you and serve you through life as a spiritual being inhabiting this intelligent organism that we call a human body. Interestingly, you realize that your body is far more capable than your mind in determining what is best for it. But listen carefully to both. As an aware being, you must learn to be in tune with all of this. This is a unified quantum process as I explain in detail later. It is not confined to just people. It is a part of our gift of life that we share with all plants and animals. However, while I will not get into it here, it even carries over to machinery... To engines and electrical components as vibratory elements that are responsive to the quantum vibrations of this Universe.

Playing this Game

This AHP that I outline is a unity, while many believe that they can do this piecemeal, it is not going to work nearly as well (if at all) at that level. In fact, one of the reasons that people die (and I outline many others herein) is: As we age we can actually can become too knowledgeable and without the sound spiritual foundation given us at birth, we become dangerously out of vibratory balance. In the quantum world, just as in these less refined levels, there is a vibratory balance that must be maintained. If you are not within those limits, the restart button is pressed (called death). It may sound harsh, but this is the reality. There are no accidents. Spirituality is thus about playing within these simple rules. Learning how love affects everything and listening to the reflections (vibrations) of those actions is the key. More on this in my other books, but this begins the story.

4. Levels of Information

First, some thoughts on critical thinking and levels of knowledge that I hope you find enlightening. They are, of course, either obvious or intuitive. The idea is to put you in tune mentally with your body and the higher levels of beingness (self):

Before we get into the nuts and bolts of wellness, let's first discuss the basis of how we make decisions in the process of mentally gathering the various sources of information into our AHP. The first concept to discuss, then, is knowingness and awareness. How do we decide what is right for us and what to do next? Herein, I give you my spiritual interpretations. Find your own and draw from them in ways that fit you best.

Empirical vs. Rational

Our discussions in the Yahoo Groups would often boil down to the science behind the program or protocol, versus our personal experience with a particular food (or chemistry). I'll speak of this in the past tense, but they are ongoing today:

Creativity Is Key

In rare cases, there were discussions regarding the tribal nature and origins of an herb. These arguments could become heated at times. They, also, could become very personal. Occasionally, my friend Duncan could send some people scurrying because they were unscientifically based.

I generally stood back from these discussions because my opinion has always been that even though the science is important, it is the creativity that is the key, and this really occurs above and beyond all of the commonly accepted mental levels that science accepts as sacred. In fact, with rare exceptions, to me, science today has become a total sham that stands between us and the truth.

Listening

I realized this years ago in my own field, when I learned to shut out everything and, basically, go into a trace state: When are aware that we are here when the inner guidance takes the reigns. Then all we need do is pick up the crumbs and apply them. I will discuss more on this later, but we all do it or have that ability, even if we have never allowed ourselves that degree of freedom. The viewpoint that allows this is what I refer to as "Soul" herein. It is our all knowing pair of eyes. It is the real you and it can dissociate with the organism that we call "our" body fairly easily if we allow it to occur... And we all do this, at least on occasion. The lessons here teach you to learn to do this knowingly with awareness and more often.

What Works

Now back to the mental level: With the above in mind, my concern is first Empirical... Does it work? Do I have first-hand knowledge of it or is that level even available? From here we can move into the nuts and bolts and possibly become totally aware of the Rational side... or what causes it to work. Knowing, objectively, "why" gives us a different kind of overview on this physical plane of existence. Both the rational and the creative (intuitive) are required in wholly achieving this first-hand knowledge... this is the bottom line that we are searching for, as enlightened beings. This bottom line may or may not be firmly grounded in science, but that is not the point here. We have no one to satisfy but ourselves in this process and if it works for us, that is all that counts. We are not running for class president and there are no constituents to appeal to. It is just you that you must answer to. So do this in the very best way you know.

Below, I begin with a method of gathering physical knowledge, a part of the key to knowing what to build on spiritually:
So what are levels of knowledge and why are they important?

First-hand knowledge

As discussed above, this is the highest level that we can personally receive anywhere, which is why it can be used it in a court of law. It does not matter what a scientific paper or studies reveals, for instance, in the case of some dietary food that we are considering, if that food does not in some way help us Thrive, it is useless. Furthermore, if we discover a food that actually does make us Thrive, we have something worth more than its weight in gold. The HL MSM protocol has been that for me. Therefore, this protocol carries first-hand feedback in my case and none of it is fear based. That is, there is no doubt that it works and no one can ever take that from me (or you). Still, mainstream medicine will refer to it as anecdotal because they have no overall studies. When you gain such knowledge, should you care if someone else calls it anecdotal?

If you follow the advice herein, which is essentially all 2nd-hand and 3rd-hand information to you, and it heals you, then you also have achieved that 1st-hand, highest level. As with me, you now own it. Again, no one can take away a 1st-hand experience from you, no matter where they occur whether on a mental or spiritual level. Finally, one must be spiritually bold and creative in order for them to occur. If they concern the AHP, you can add them to your basket of goodies. A final point along these lines: Creativity, itself, may be subjective to others, but it can be extremely objective to you, the viewer (5th edition).

So what are the less valid levels of knowledge and... are they important?

Second-hand knowledge

2nd-hand knowledge is logically the next level on the ladder and medicine refers to it as anecdotal... And it is, even to you. However, this is still valuable to us most of the time. There are many times when this is all that we have, but always keep in mind the source.

Improving the System

Application: For instance, When you tell someone about the *HL* MSM protocol and the next person discovers that it works for them, the protocol benefits and is raised to the highest level. More will begin to use it. It does not increase in value to someone who has 1st-hand knowledge, because, to them, it has already been proven. There is a perceived higher level, nevertheless, and it is very useful for advancing systems and protocols... and these people generally improve the system as it advances.

.

Lower Knowledge Levels

Obviously, beyond the 2nd-hand-level... the information decreases in value considerably. 3rd-hand would be when it works for someone else and they pass it on, in turn and you get wind of it. This is probably the lowest that we should ever consider ourselves when evaluating a system for anything. At this point and lower, the information can become confused and even harmful. Its like the parlor game... this incorrect information is often out of control and especially at the 4th-hand level. Of course, this is why we write things down, but also why the internet is full of misinformation along with the good stuff. Still, we can investigate anything today.

The Whole Story

Finally, when you hear me (or anyone else) say, They say, Experts tell me, or Scientists have proven, you know that we are working at a very low level of knowledge (attorneys call it hearsay beyond 1st-hand). At this point, you must either take their word for it (on faith) or simply discard it as useless information. At the faith or lower anecdotal level, if a healing concept or protocol is completely counter to common knowledge, many will toss it aside after reading it and could miss out. The point is that we have sources to verify facts today that are quick and efficient.

We have an incredible library of information on the web that can turn the above anecdotal information around quickly. Remember, the idea is to improve our source level... to eventually make it 1st-hand information. So the bottom line, in your search for the best solution, is to finally get 1st-hand knowledge, not heresy. This virtually makes your body a test-bed for wellness. It sounds kind of risky, but if you think about it, it has always been the only "real way" of achieving a high level of wellness. However, we have never had this level of capability and speed in the history of man to keep us well. This is virtually how the entire AHP has been formulated and spread... and you are benefiting.

Creating Requires Every Level of Knowledge

Arguably, faith is considered the lowest level of understanding or discovery on the scale of objectivity as opposed to scientific knowledge. However, creative people, those paying attention... who come up with new things, visions, and products... must have a high degree of faith in their <u>instincts</u> and be willing to work on every level at once.

An Example

Architectural Design is a good example of this: A building must in some way present a fresh viewpoint, but it also must be sturdy and stand the test of time, be structurally sound, and function on many levels. So a good building must cover the spectrum from intuitive knowledge to the cold objective mathematical considerations like structural soundness. My premise here is that, to be creative, you must use all of these levels, even those higher levels of knowledge unknown to science, in your search for your AHP.

Skewed Science

In today's world, face it, everything is generally driven by money. In this process, government or corporate money pays for what was once high level scientific knowledge. Thus money strongly influences much of what we now call science. However, despite this glaring short circuit, we are still told that the science and scientific reporting are always objective. Incredibly, even government-backed corporate studies are, nevertheless, still referred to as scientific. In fact, "Skewed Science" is worse than none, because it serves as a false authority that actually causes hardship and can even result in death on occasion (actually, millions according to Dr. Mercola).

Listen to Your Body

The idea, of course, from a Pharmaceutical Company's view is that they convince you to believe that they carry a higher level of authority than you can possibly attain and, therefore, that you need their product. Certainly, there are instances of when this is true... but one, for certain, is often not: That is, when you have gathered first-hand knowledge that totally disproves theirs. The point is that it is your body and you can hold the highest level of authority, not them.

Anecdotal Evidence
Interestingly, in a video interview Jon Barron **http://www.jonbarron.org/**, speaks on blood transfusions. Here he cites a study that proved that transfusions (which have been an accepted medical practice for more than one hundred years with no studies) are dangerous because the Nitrous Oxide is depleted from the donor blood by the time it gets to the patient. As a result, they can cause heart attacks. No one ever knows...

Informed Anecdotal Level Jon further points out, that mainstream medicine mainly operates at 85% word of mouth (mostly 4th level or 15% proven... or at an "informed anecdotal level." 3rd level). From here, he asserts, alternative medicine is held to an absurdly high 100% (1st level) requirement by both the AMA and FDA. Jon also points out that "studies" can be designed to prove any side of any story and prove any point, which is my above underlying story (5th edition).

My point again is that the 1st-hand level of knowledge, as just outlined above, is still the best that we can do, overall, on this earth plane. Furthermore, it is far better than anyone else can access, because it is specific to your body.

5. Your Body is a Quantum Machine

So why are we able to work beyond even the intuitive level? Those who have been there in full consciousness could never explain it, but I submit that it really does happen and it occurs because: "Your Body is a Quantum Machine."

Much has bee written of quantum mechanics and most of us really have no idea what it means.

Watch the following Brian Green Video:
https://www.youtube.com/watch?v=CBrsWPCp_rs
Where things appear in more than one place at once. As they say, they basically contradict our physical laws (Newtonian), yet are always correct. Unlike those few bold scientists who pursue quantum mechanics and beyond, the supposed rational theories of today's scientists are mainly focused on and fiercely protective of the world view handed to us by Sir Isaac Newton, the genius who set up most of what we believe today as science (including medical science). They are often contradicted, however this is never the case with Quantum Mechanics (QM).

So the interesting thing here is that not only is Quantum Mechanics so predictable, Quantum Biology is equally so, and our basis for modem medicine is generally incorrect. For a thorough discussion on this read my book, "Pretty Fine Sex is Spiritual." In it I show you a plant experiment that speeds up plant growth 400%.

Newton Was Wrong

My premise here, and that of others like Astro and Quantum Physicist Dan Nelson, **https://www.youtube.com/watch?v=7hNW7qxIMzg** who has applied it to the human body, is that Newton was wrong in many respects. His science left out an entire level of knowledge that we carry with us as living organisms at the cellular level. Our bodies are quantum machines that today's medicine often attempts to repair while modeling our bodies after cars or airplanes, supposedly working under Newton's rational scheme. Look at my car test videos on You Tube to see how it is all wrong.
https://www.youtube.com/watch?v=YFakNUEQoNM

The current Newtonian theories are flawed and full of contradictions that were often simply disregarded. Our industries, especially the pharmaceuticals are entrenched and highly invested in them, but they must eventually give way to what actually works based on quantum physics and fractals Even though the fast cures demanded by society today can work short term, drugs are band aids and, as Dan Nelson says on his video, are poisons. For a thorough discussion on these terms read, "Pretty Fine Sex is Spiritual."

Most scientists who study this new (old) science maintain that all organisms function on the Quantum (microcellular often) Level as predicted. This means that physical science observations (Newtonian) at the microcellular level are inaccurate. Beyond that and interestingly, I only discovered, first-hand, that even our diagnostic machines do not function as claimed on this level, strangely, even though they are the basis for all mainstream diagnostics (5th edition).

In truth, this whole Universe is a Quantum Soup and we are just a part of that. Therefore, our physical observations are pathetically inaccurate and our world is one of nearly total delusion just as the early saints and spiritual masters have been trying to tell us for millennia. Relative distances and times are mental illusions when one does Quantum Healings. Therefore, healings are often instantaneous, and even when done with the vibrations that the Drs. Dan Nelson and Bruce Lipton teach at the subconscious level (5th edition).

Now, with this new certainty, the early geneticists like Dr. Bruce Lipton has surrendered completely the genetic determination sworn by the mainstream. Today, Bruce teaches that on the sub-conscious level, any person can heal themselves (not always true, but often). Please do not confuse this with the "power of positive thinking," This occurs as Bruce demonstrates here: **https://www.youtube.com/watch?v=YackvFSlDQk**

You are Just Amazing

Beyond the above, I submit that, were it possible to actually measure your body's energy efficiency, in whole, you would discover that your body achieves "over-unity." By this I mean that under certain conditions, it actually borrows energy from the vacuum of space (the quantum realm). It does this according to the various inputs that you control as a higher spiritual entity. Additionally, I submit that, by this, it is possible, to actually visualize our bodies at this higher level and, in turn, create an organism of greater abilities than any that have ever lived. That is, as we make this visualization a part of our very own **AHP** and allow this unique body that we inhabit as Soul (which I further submit is actually working at a higher level than even the quantum scale), we can achieve superhuman abilities. Finally, in doing this, we pass on these changes to our offspring as Epigentics teaches. Thus, if this is so, we blow Darwinian Theory out of the water. It has no application in these realms. Also, this explains many of the past discrepancies (shortcomings) associated with Darwinian theory (5th edition).

As I suggest above, most physical scientists are not concerned with these levels because they cannot easily observe them. All the while, Dan Nelson and my friend, Dr. Rich Price consistently see healings with very accurate repeatability, thus proving this theory is going to take awhile. Repeatability is, after all, the only merit that you can assign to any science and, thus far, quantum science never fails (5th edition).

Nevertheless, in my 1st-hand experience with quantum machines (cars here), quantum events are interesting in that they may be reproducible fifteen times, then appear to go away, only to reappear for the next twenty, but there is no error. This drives Newtonian Scientist's out of their minds, but the events that I describe above occur with common regularity throughout the Quantum world. Our minds can influence the performance of both machines and biology, but I submit that these occurrences generally go unnoticed in the current world of science. The key to real resolution rests with methods that consider our influence in outcomes, not in negating them (5th edition).

Furthermore, we actually witness examples of this involvement in athletic events, often, but pass these accomplishments off. Thus, these are common occurrences. The fact is, the athletes, as quantum machines, are influenced by the minds of others. Furthermore, crowds can interact as a group to reinforce them. This all relates to the quantum level, as

I suggest. Read Pretty Fine Sex is Spiritual to get this (5th edition).

Before 1954, the idea that anyone could run a sub 4-minute mile was absurd. But that year, Roger Bannister proved the "impossible." Today, high school runners cross this threshold. The common theory is that we are better fed. Actually, the reverse is

undoubtedly true if one compares the nutrition of that era to today's. What is true is that everyone knows that there is no longer such a sub-conscious barrier. So where would Bannister score today if he could see today's runners and relate (5th edition)?

"What is now proved was once only imagined," per poet William Blake and, more often than we could even imagine, we may have witnessed events that exceeded Newton's rules of physics right before our very eyes. This puts a whole new light on athletic events when you keep this in mind and goes well beyond current science.

It Happens Everywhere

A second example of *"amazing" phenomena which we do not "see" occurring before our eyes* was recently related to me by a friend and engineering consultant, Gary Bossard when be noted that occasionally we see a flock of starlings... *maybe a thousand birds flying...* totally in unison, *and then they all* wheel as one bird... with no time in the movement. Gary submits that there is an inner quantum level conversation going on that exceeds words or even rational thought. Do you disagree?

So my premise is that we all carry with us this same wonderful capacity *to function on levels we have not yet achieved.* But, notice that I am not saying, "more than *we can* imagine," because visualization is the first step to breaking into the sub-conscious realm and the barriers that it holds. Dr. Lipton stops at the subconscious level, but there is no end to the spiritual realms and they continue. (5th edition). .

It's the Observer Stupid

Yes, Quantum Events are not as repeatable as Physical Science would like in certain ways, but in others, far more so. This has more often than not been the rub till recently when we discovered that the individual observer (the most powerful of all Quantum Machines), can actually affect a Newtonian scientific experiment. This is best answered

by the two-slit experiment discussed on (and on numerous You Tube videos):
http://answers.yahoo.com/question/index?qid=20120122224633AA50B0A

Here, repeatedly, the observer alters the outcome. This can and does occur in Newtonian science's time honoured double blind testing. Basically, as Quantum Beings, the bottom line is that we can change virtually anything under the correct conditions. This means that, just as Jesus reportedly walked on water, so can you, but this level of physical "interference," if you will, is counter to natural occurrence, disruptive, unbalancing, and thus very uncommon, for good reason. Still, what looks like magic in terms of healing is apparently fair play and not nearly as difficult or unusual when correctly achieved as commonly assumed. Thus, at the Quantum Healing Level, blind men occasionally can have their eyesight restored or they can get up from a wheelchair and walk and this is not magic. The individual must always do the work. We self-heal, no one can do it for us. Once you allow the quantum in and witness it, everything changes along with it.

Skewed Reality

Dropping back to the physical level, even if Sir Isaac Newton were totally correct in his methods of reasoning from every perspective (and there were no quantum science): Newtonian science is not being applied effectively. By the way, Boolean algebra and quantum mechanics, tools that allow for a whole new perspective, were completely foreign to Newton. My premise, however, is that, given Newton's narrow view of (objective) reasoning, so much importance is and always has skewed reality.

I submit that anytime real advancement has occurred in any field, the higher levels of knowledge *were* part of the process (from intuition up), even though they went unrecognized. That is to say, for anything to be totally successful, these higher levels (beyond "normal" rational reasoning) take place, but few are aware of them and they are, thus, disregarded.

The Bottom Line

When you are reading this or evaluating anything: the bottom line is that we need to look at everything from the widest viewpoint possible. Thus, we must always give consideration to the level of knowledge being presented as discussed above. Finally, visualize as well, the attitude of the person presenting the information and add that to the mix and balance it. This all adds up as we progress in the school of life. If we are applying the tools we have learned, we continue to graduate to higher levels of spiritual beings as our perceived limitations and barriers disappear.

Reduce Your Steps

When we become the person carrying the knowledge 1st-hand, all the better. In any event, we all seem to understand the above, at least subconsciously, because, as spiritual beings, whether consciously or not, we are working on all levels. And, yes, on occasion, we simply must take things on faith, or the "Informed Anecdotal Evidence," just to begin the adventure. Occasionally, this is the best possible way of getting to the higher levels of understanding.

In conclusion, reduce your steps to 1st-hand knowledge to a minimum whenever possible, but take your time. If this sounds like a paradox, it is. Spiritual adventures must allow room for a pinch of the irrational (referred as, "Letting Go") in the mix, if creativity is going to be the outcome.

Time Lines

When it comes to evaluating the various foods and supplements presented here, 1st-hand experience should always be the goal in every case. However, often the time required to see results (feedback) is excessive. Therefore, we must occasionally rely on the experiences of others and accept their results on faith as a first step. The dilemma, of course, is that even if their results are correct for them, they may not work for us and our individual findings.

The next dilemma is that foods and supplements, especially, are generally very inexpensive compared to drugs. There is simply not enough profit margin in supplements to finance good repeatable studies. Since, for example, no one can patent Moringa Leaf Powder—even though it may heal more people, in a shorter time, with no negative effects, than would any targeted drug. Since profit drives the pharmaceutical industry, this is not about to change anytime soon.

Advancing Our Experiment

In the case of the ***AHP***: if our ***AHP*** were patented and double-blind studies were used, it would more than likely take 50 years to conclusively prove the effects—based on

Newtonian sciences. However, if we can get a number of us on the ***AHP***: when we are well past out grandparents' age and still playing a solid game of tennis, or whatever, believers in the system will be coming out of the woodwork. I am asking your participation here in helping me provide this proof. This, then, can be our own study.

6. Golden Coins

Below are some simple Timeless Concepts that can help keep you well and *function optimally* on all levels:

Don't Hold onto the Past

The corollary to this is: "Live in the Moment." Of course, this adage has been repeated by virtually every spiritual master that ever lived. It is simple, but oh so hard to follow. When you cling to the past, you hold that consciousness within your physical body (and the way you interact with all others). This invites illness. Even if you did everything else perfectly, an ill thought must manifest itself at a later point and it can easily manifest as a disease. Notice the self explanatory term, "Dis -Ease?"

Take Responsibility

When a past event manifests itself in an illness, most end up blaming the cause on extraneous issues, "Our son brought the virus home from school." Worse, if you hold onto that level of consciousness very long, it could be manifest in such serious diseases as Cancer or MS. Or, taken to the extreme, sudden death could occur due to spiritual exhaustion. An example of this can occur when a person's spouse dies and they quickly follow. This is often described, medically, as a "coronary infarction."

I submit that few ever die from being too happy, but holding onto sadness over a past event can lead you to an early grave. As noted above, less dramatic examples occur every day. A one-day virus could be just an ill thought escaping. If "not letting go" is the cause, then you caused it.

Take Your Time

Since it is all an illusion anyway, you have infinite time. When you make changes in your diet or the way that you are doing anything, slow down, and be certain that you have a clear grasp of what it is that you are doing. Visualize the outcome, and evaluate it on all levels. Be totally aware of conditions... take it slowly. Try to observe *how* each change *affects you* individually.

Making two changes at once means that you may never know which was effective... (keep this in mind and I'll remind you again). Taking two new supplements at once, negates any possibility of objective outcome evaluation.

Visualize Outcomes

Life is an experiment, your experiment. Again, avoid focussing on the past and coming events as much as possible. Stay on point and keep score. This is difficult at times, but you really do have "plenty of time." By visualizing outcomes and projecting them, you begin to spiritualise the mundane, you begin to take charge of your future. You are no longer a slave to anyone else's vision and your dream is yours to fulfil. This takes your entire life into this grand experiment Enjoy the spiritual journey.

Finish the Job

I spent most of my youth (note that I avoid using the term Formative Years, because we should always be "re-forming," On my *gr*andpa's melon farm in Southern Indiana. I absolutely adored my grandpa and grandma. They had nine kids and, two, my uncles, Jack and John, were only four and five years older than I. They served as friends and, often, in those days, were my closest mentors.

Riding *the Farmall M*

When I was about 6, my older of the two Uncles, John (11 then), asked me if I'd like riding with him on grandpa's new Farmall M. Now, a Farmall M was quite a piece of equipment for anyone to have in those days... and I was delighted. We set off at about noon to mow a field of barley for the neighbouring farmer. Neither one of us really understood how big of a project this was, but we were both anxious to get going... we were having fun. What a day!

Seven or eight hours had passed: We had mowed the entire day. Daylight was fading quickly. By that time, I was ready to go to sleep. And I was starving. So was John. We still had at least an acre left to mow and those rows were looking much longer. I begged for him to stop so we could go home. But John explained in his responsible 11-year old way that he had agreed to mow this field for twenty bucks and, that we had a commitment. Twenty bucks, in those days, was at least three or four's days' work at turning melon vines for an 11-year old. I understood. Looking into his eyes, I realized that it was far more than the money that we were discussing here. There were several things going on at once: responsibility to the neighbours, sealing an agreement, and keeping one's word were a few. But it was clear that John was going to finish that job, even if we both ended up falling off that M in the process..

Making an Agreement

Today, when I agree to doing something. I think back to that M tractor ride. The bottom line is: When you agree to something, finish the job, and always do your best, no matter what the circumstances. There is no getting around that lesson. I ride that M first-hand.

Your Agreement

So how does that story relate to wellness? You have agreed to live this life doing the best that you possibly can, even if that agreement is forgotten or unclear for now. You were given a tractor. Hopefully, it was a nice Farmall M, but whatever it was, you are here to do the best job that you can with the equipment issued. My advice is drive until you fall off of that seat. If you do, you will never regret the trip. According to how well you do it, you will reap the rewards of that 1st-hand knowledge. Most are not aware of it, but you carry that ride with you through infinity. Nothing is lost spiritually. You actually do take it with you, even as the equipment changes.

My Heroes

Many well-meaning people love seeing us as victims and the masters will warn us to avoid such people. They seek your total dependence and, in turn, they grab your freedom. Fortunes in the advertising industry are made based on this formula. If Apple's I phone is required for communication in your business, your livelihood depends on it. They get to drive the M and you are hanging on for dear life trying to stay awake. When a corporation does that, they can expand and gain even more control. They take the driver's seat and you go along for the ride. So visualize yourself in the M seat... with the best equipment out there and stay in control. No one else drives your M.

Below are two people that I admire who bucked this above trend:

Mrs. Alexander lived across the street from me for almost twenty years. She asked me to repair her screen porch for her once when I first met her. I quickly got my hammer and staple gun and had it fixed in maybe an hour. She gave me two dollars in payment and I lovingly accepted it. She was a very independent woman and never seemed to get sick. Then, one day there was a small range fire caused in her kitchen when she burned some beans. The volunteer fire trucks came, but they had little to do. It was already out when they arrived, but there was plenty of smoke. Within a week, they took her away. She was in her 90's then, but clear and alert.

She and her grandson Steve both fought them, but she was placed in what we used to call an "Old Folks' Home." Her house was sold to pay for her support and everything was ripped from her including her independence. Steve visited her, 50 miles away, as often as he was able, but she was gone in maybe three years. He and I still discuss it when we meet over coffee at our favourite local restaurant called the M&H. The local watering hole as my old friend Walt calls it.

Steve now lives down the road in a house trailer that he loves. It was donated to him. He contracted a rare disease at a very young age that left him with malformed short arms, hands and legs, but Steve never let this condition make him a "victim." He is wheelchair bound for the most part, but manages to somehow move his collapsible wheelchair in and out of the local M&H restaurant in his pick-up truck. He parks his truck in places where people with just shortness of breath and HC tags would never go. Steve is unbelievable in manipulating that wheelchair given the "tractor" that he has been dealt.

That said, my friend Steve has always had problems with arthritis (and worse). Last year, I bought him a bucket of MSM with my instructions. He is independent, but always grateful. No one out there is more deserving of optimal health than Steve. That was about a year ago... And, yes, he has improved. M&H closed for a couple of years because Aunt Mary, who ran it, died. But fortunately one of the regulars picked it up. We all stopped over to see Steve, though, at his house trailer—even when M&H was closed.

It's good to see that we are back in business though. There are very few places left in the country like M&H....where Golden Coins are minted.

Rules Are Made to be Broken

So, with the above rules in mind, proceed with confidence on your journey. This path, the **AHP**, if you adopt it, allows you to control the throttle. It's your M. It can lead you to new levels both as a superior organism and a conscious Soul, a Soul completely aware of the intuitive subconscious and the worlds beyond. Learn what the rules are and have been and why. Experience it first-hand. Read and research. But break the rules when you get the urge (feel compelled)... all along... do the work. No one can do this for you. Thanks for listening to my heart speak out to yours.

7. HOW SAFE IS MSM?

So back to the more mundane, we discuss the critical aspects of the first star in our (AHP) Anti-Aging & Health Program:

Before we get into discussing HL* MSM *itself, and related material, the below MSM descriptions etc. were copied from an organic chemist's findings and Byron Richard's site. *This was taken (snipped), except* ***for my comments****, from the following article by Byron Richards at:* http://www.wellnessresources.com/health/articles/msm**:**

- MSM is considered to be one of the least toxic substances in biology, similar in toxicity to water.
  Note: I have read articles by other experts who say that it considerably less toxic than water. Then, they go on to say that it's the least toxic of all chemicals ever observed and, possibly, on earth, but this serves my point. While it may give a very few, some stomach discomfort, at least in the

beginning, its not going to harm anyone to any degree, no matter how high the dose, once you have gotten past the initial start-up period.

- When MSM was administered to human volunteers, no toxic effects were observed at intake levels of 1 gram per kg of body weight per day for 30 days.
- Intravenous injections of 0.5 grams per kg. body weight daily for five days a week produced no measurable toxicity in human subjects.
- The lethal dose (LD50) of MSM for mice is more than 20 g/kg body weight MSM has been widely tested as a food ingredient without any reports of allergic reactions.
- An unpublished Oregon Health Sciences University study on the long-term toxicity of MSM over a period of six months, showed no toxic effects. More than 12,000 patients were treated with MSM at levels above two grams daily, without toxicity ***(End snip)***

Therefore, the lethal dose by his computation amounts to approximately 1620 grams or 100 level tablespoons for a 180-pound man, in this man's opinion... And the above is conservative from other reports that I have read.

The bottom line here is that there are far easier ways to commit suicide than taking MSM... and no one is going to kill themselves with an overdose, so take as much as you need, once past the start-up period.
Also, read this ref: http://www.naturodoc.com/sulfurstudy.htm

8. MSM and the Detoxification & Start-up Period

Below is my summary of the nuts & bolts of MSM and hints on the critical start-up period:

- Many people who start using small amounts of MSM notice very little effect at the onset, or they may experience slight detoxification symptoms during the "start-up period." But most feel the healing process begin immediately and with no negative effects period.
- In fact, a good many reading this, have already taken as much as a gram a day of MSM in conjunction with other joint supplements. While these were, likely,

fairly ineffective in terms of curing their condition, most likely, it has readied them for the *HL* MSM Protocol, so they can effectively avoid any start-up period as I did. Some may choose to jump into the *HL* MSM Protocol immediately. Most of these will see immediate improvements with no detox, period, and no negative effects

- The prudent thing is to, again, take your time. There is no contest here and no one warned me about possible detox effects. So I was fortunate.
- While you may be ready to go, its best to go slowly... But it is your body and there is a form of self-regulation going on. If you fail to heed the above, your body will speak.
- Start-up period symptoms (Herxheimer Reaction) may include mild forms of diarrhea, skin rash, headache and fatigue. Fewer than 20% of users of MSM may feel moderately sick in the first few days of using MSM. After one week, all symptoms commonly disappear.
- It may be a small consolation to know, that the stronger the detox symptoms are, the more toxicant stored in body fat and the more MSM is needed for purification.
- Whereas toxic metals can be fairly benign when stored in body fat, they can be quite poisonous in your gut. Keep that in mind.

Helping Out

During this period, you may be possibly releasing heavy metals, like mercury from fillings into your system. If you have any of these such symptoms, you want to excrete them ASAP, back off, and slowly go back. One means of speeding this period up is to spend a great deal of time exercising... and basically sweating, by any means possible... a sauna or hot tub would help this. There also helpful detoxing supplements like ALA, Taurine, and Chlorella that can help.

MSM Effects (4th edition)

MSM will successfully relieve pain, in most cases in minutes, while working toward a long-term and permanent cure of the following:

- Personal injuries from accidents

- Lower back pains
- Osteoarthritis (OA)
- Rheumatoid Arthritis (RA)
- Fibromyalgia
- Whiplash
- Sciatica
- Headaches and Migraine
- Muscle aches and Strains...
- Bursitis
- Tennis elbow
- Repetitive Strain Injuries
- High Blood Pressure
- Cardiovascular Disease
- Cancer such as Mesothelioma caused by poisonous materials (so far)
- Carpal Syndrome

And most importantly, as related to most of the above problems

- Liver Diseases and their consequences as outlined in detail herein.

And wipe out entirely, in many cases:

- Sclerosis of the Spine
- Burns
- Scars due to burns, operations, accidents, etc
- Type II Diabetes and Type I to some degree*

*Type I diabetes falls into a similar category, but once contracted, it is probably forever, as I understand it. Give me your feedback please. We just do not know yet. but I am hopeful.

MSM Mechanisms MSM impacts against pain by the following mechanisms:

- Improves liver function... This is addressed below in further detail, but this effect alone drives nearly every other characteristic described in this list as you will see.
- Improves the permeability of cell membranes, thus helps you hydrate more effectively (see Ch 37 MSM's Secret for more on this)

- Methylation provides positive mitochondria and telomere effects (See Ch 9 & 51) while stopping pain better than any analgesic or aspirin possibly can (also, consider: calcium causes muscle contraction while magnesium relaxes them. See Ch 21 on Cacao for this).
- Analgetic: blocks the transfer of pain impulses through nerve fibers (C-fibers)
- Increases Electron Transport as explained in many places herein. A key health factor.
- Blocks inflammations and inflammatory processes thru "Methylation."
- Converts Vit B1 and Biotin for energy production.
- Contributes significantly to Keratin production for better skin, hair, nail, and hair.
- Enhances the activity of cortisol, a natural anti-inflammatory hormone.
- Improves the uptake of nutrients, vitamins, and minerals in general. Thus, higher energy levels (in addition to adding directly to improved methylation).
- Helps eliminate waste products and excess cellular fluids through "Sulfation."
- Dilates blood vessels, enhancing the blood circulation & lower blood pressure.
- Speeds up all healing processes both thru methylation and sulfur contribution.
- Muscle relaxant... Chronic pain is aggravated by muscle tension.
- Improves all natural defense mechanisms in the body by regulating prostaglandin metabolism, and regulates the formation of antibodies and immune complexes.
- Burn scar repair: MSM ointment slows down and restores cross linking in collagen. Cross linking is a natural process in scar formation, causing hard and often painful scar tissues. With burn scars, large surface areas may be affected, this may lead to chronic pain. DMSO is a trickier, but even more effective pathway to help with this also (discussed later).
- MSM contributes to Liver Health by supplying the sulfur containing amino acids Methionine and Cysteine while boosting Glutathione levels significantly.
- Healing scar tissue makes the skin more flexible and stops pain.

Watch Dr. Mercola's video on this at: http://products.mercola.com/msm-sulfur-supplement/?e_cid=20130301_DNL_banText&utm_source=dnl&utm_medium=email&utm_content=banText&utm_campaign=20130301

Also, Read the following discussion on Sulfur: http://www.acu-cell.com/ses.html

9. Telomeres & Aging

***Below we continue with more of the AHP, with discussion of the current scientific theories of the causes, effects, and measurements of aging and possible methods of limiting it at the cellular level*:

Telomeres

Telomeres are microbits of protective DNA. First discovered ca.1970 by Russian Alexei Olovnikov, they can be seen at: http://en.wikipedia.org/wiki/Telomere. Telomeres appear at the ends of chromosomes and are known to shorten with cellular aging. Telomere length, therefore, has been suggested as a marker for biological aging, chronic disease risk, and premature mortality. If so, a long telomere indicates youth, but there is plenty of room for more study here before cause and effect are separated as a certainty.

Telomere Study:

Researchers in Helsinki, Finland measured leukocyte (white blood cell) telomere length in 1,942 men and women enrolled in the Helsinki Birth Cohort Study of subjects born between 1934 and 1944 at Helsinki University Central Hospital. Self-administered dietary questionnaire responses were analyzed for the intake of fats, fruits and vegetables.

Telomere Length

Shortened length has been linked with: coronary heart disease, heart failure, hypertension, diabetes, and some forms of cancer. However, this recent study associated the shortening with a reduced intake of vegetables in women and fruit in men. (December 2012 issue of the European Journal of Clinical Nutrition).

Hey, I did not design the study... Why the different results in the two sexes? Is this the Jon Barron effect? *Read more on Telomeres by Byron Richards at:* http://www.wellnessresources.com/health/articles/how_nutrition_makes_anti-aging_possible_secrets_of_your_telomeres

Is It A Marker?

Finally, this particular study drops back to warn us that Telomere length is more of a species observation... and that telomere length is not necessarily a true marker for aging. At this point, contrary to some earlier reports, if this is true, the study sounds negated However, there is reason to believe that we can influence telomeres with just our attitude as previously discussed. Also that MSM is a key player as noted below with methylation and thus Telomere length. So do not give up on this as an antiaging effect worth pursuing as we read on:

10. Methyl Component of Telomere Function/ B-12 (4th edition)

Continued discussion of the current scientific theories behind the causes, effects, and measurements of aging and how MSM, Vitamins B12, B6 and Folic acid may impact:

Methylation

Healthy Telomere function requires adequate "Methylation" in order to turn genes on and off... to regulate properly. Methylation is the chemical process of donating a methyl group (one carbon atom bonded to three hydrogen atoms (or CH3), univalent hydrocarbon radical, to the genetic material of the telomere. This epigenetically marks the Telomeres, making them behave properly. http://www.answers.com/topic/dna-methylation (Also watch Update videos)

MSM (Methyl-Sulfonyl-Methane)

MSM obviously contains a dual organic methyl component that is bioavailable in the three Methyl Cycles. For your Telomeres to work properly, just as your car needs ethyl octane to prevent knock, you need methyl donors. The primary commonly addressed methyl_donor for this purpose is SAMe. SAMe is directly derived from a **methyl donor**. Organic Sulfur (**Sulfonyl**) and is thus a key building block in SAMe. SAMe also requires the B vitamin building blocks: vitamin B12, B6, and folic acid. B12 thus plays multiple roles in supporting telomere gnomic stability. If you are inclined, read: http://www.biomedsearch.com/nih/Telomere-dynamics-genome-stability-in/18160783.html However, for our purposes, Gnomic Instability = Aging. Do you see where this is going? Mainstream science does not get HL MSM yet. We make Sam-E, but we can supplement with it also.

B Vitamins

Should you supplement with these B vitamins? Good question. The answer is, first, always get nutrients from your food when possible, but most nutritionists believe that organic B12 (methylcobalmin) is important enough to supplement for many reasons beyond the fact that it is a key component of SAMe. Some common indications of a deficiency in B12 include: chronic cough and a vegan diet. They then move to: growth retardation in children, dementia, heart disease, anemia, among a host of other devastating conditions. The warning here is that when you short circuit energy flows, the results can be dramatic. Niacinamide (B-3) also has some beneficial effects as does organic B-6 (P-6-P. I include these in my own transdermal cream that I make for myself for this reason. Do your homework, though. (5th Edition)

A Word on Supplementation:

If you decide to supplement any vitamin, always make sure that you are getting an **organically based source**. As Dr. Wallach (later on him) would tell you, artificial vitamins are never a good choice. My warning takes it a step further: artificial vitamins can actually create deficiencies. Also, few artificial vitamins have ever been critically tested. Finally, they can fool your body into seeing them as ghost nutrients that are not actually there. If you are new to supplementation or unsure, consult with a nutritionist. Few vitamins, even the most common, like C and E, make my list, since I eat high grade greens grown in my own garden when possible with no artificial fertilizers or insecticides. Also, I eat plenty of whole citrus fruits including organic lemons and limes (low sugar levels), when possible, skin and all (more on that later).

Vegan Diet

'Organic sources of B-12 include mainly animal meats, but also organic raw milk and eggs... but also, fermented foods to some degree. The problem with just vegetable sources (vegan diet) is that it misses a key analog and can cause problems that can actually make for a B-12 deficiency (and worse according to Dr. Mercola). Read Dr. Mercola's articles on B-12 at:
http://search.mercola.com/search/pages/Results.aspx?k=Vitamin%20B12

Fermented Foods A similar Vitamin to B-12, K-2, can also be derived from fermented foods. Read about it at: http://chriskresser.com/vitamin-k2-the-missing-nutrient
High on my list, K2 (MK-7 form) is especially effective against prostate cancer, but it protects against heart disease, ensures healthy skin, forms strong bones, promotes brain function, supports growth and development... Just to get you started.

Obvious Results However, do not expect the level of results from supplementation of any of these that you actually observe, short term, with MSM. Every other supplement that I have ever used is a long-term event and unless you happen to have a very high deficiency in a single mineral, you are unlikely to ever notice an immediate improvement. There are a few other exceptions like our special water as discussed later, and astaxanthin for sunburn, but this is the rule.

SAMe Also, you can actually supplement with SAMe alone, and I have; however, I have never noticed any effects from using it, same as above. If you suffer from anxiety problems, it may be a good choice for you in addition to the *HL* MSM and UWPP. In my case, I have never had anxiety problems in this lifetime, but if you have, you may notice some improvements. Personally, I have always "known" that I was not going to "fall off of the M."

Additionally, I have spoken with people who have noticed that their entire feeling of well being is enhanced by supplementing with SAMe. As far as I know, there are no known problems with using it, but my thought is that it a fairly expensive way of achieving what the AHP does cheaply.

11. Mitochondria, Anti-Aging & Calorie Restriction (4th edition)

Further discussion of the current scientific theories behind the causes, effects, and measurements of aging and theories regarding calorie restriction:
http://micro.magnet.fsu.edu/cells/mitochondria/mitochondria.html
An even stronger marker for aging, beyond Telomeres, is seen in Mitochondria:

Mitochondria are peanut shaped organelles that range in size from 1 to 10 microns in length. They are described as the "cellular power plants" of animal organisms, but are also present in all plants. They are the source of adenosine triphosphate (ATP), a coenzyme, which is the cellular transport energy used in both cellular metabolism and signaling between cells.

Normally, these organelles are beautiful long tubes. However, as cells grow older, the mitochondria become "fragmented and chunky" according to observers (see pictures online). Furthermore, and more importantly, they decrease in number as we age, so our mitochondrial count has an aging factor that is not commonly stated, attached to it and this alone may be the single greatest aging hint of all. Conversely, increases in average cell count would indicate an internal reduction in age. In our experience, a sick or dying person may have 50 per cell average and the most healthy 2000. So you can materially influence the count as can correctly employed fractals (and our water which is fractally based).

Furthermore, mitochondria are theoretically the reason that you can taste food. This writer believes that the fact that MSM cannot be substantially tasted when you swish contributes to the proof of this theory. Furthermore, with regard to earlier discussions on Quantum Effects, if those premises are correct, and the writer also believes that they are, this is likely the primary seat of quantum events in organisms, since ATP is the source of body electronics. Therefore, mitochondria are responsible for the energy that makes this communication possible. So, the bottom line would be that happy mitochondria yield happy people. Also, it follows that, as energy levels (ATP) increases cellular age decreases (read "Pretty Fine Sex is Spiritual" for a current report on this).

Mitochondrial Efficiency There is evidence, as noted later herein, that beyond liver improvement, the Biological Transport Pair MSM and DMSO actually clean up mitochondrial efficiency through methylation in a way similar to the above described telomere effect. However, this requires more study and most scientists are not well equipped to understand quantum events. So this is going to take awhile. In my first-hand experience, as previously discussed with quantum events, they may be reproducible fifteen times, then go away, only to reappear for the next twenty thus driving a Newtonian Scientist out of their mind, The bottom line then, is just be happy that they occur often and encourage it as I suggest here.

Keep in mind that the AHP effects acidosis also. But let's assume that Mitochondria are indeed cleaned up and even made younger by the AHP, because as this writer believes, this is exactly what occurs: if the above is correct, the AHP opens up the possibilities for these Quantum events to occur more easily and, thus, more commonly (assuming that they are not commonly or always occurring already).

Cut & Paste http://www.medicalnewstoday.com/articles/253172.php to read the original information edited and rewritten from Medical News Today:

Experiments On Yeast The following is an interesting related study on acidosis and calorie restriction concerning mitochondria: In the U.S., for the first time, scientists have identified key events that occur early in the aging process of cells that may explain how genes and environmental factors, like lifestyle and diet, interact to influence lifespan, aging, and age-related diseases like cancer.

The changes occurring in shape seen in aging yeast cells can also observed in certain human cells, such as neurons and pancreatic cells, and those changes also have been associated with a number of age-related diseases:

Low Acidity of Vacuoles Impairs Mitochondria

The reason that mitochondria becomes misshapen and stop working so well has until now been a mystery, but researchers point out that certain changes inside vacuoles may be responsible. Vacuoles are tiny pockets inside the cells of organisms that help break down proteins and store molecular nutrients for the cell. The equivalent of the vacuole in human cells is the Lysosome.

Cellular Acidity

This appears to be critical to the aging and function of mitochondria. In particular, the low Acidity of Vacuoles Impairs Mitochondria. Vacuoles become more basic quite early in the life of a yeast cell. When acidity drops, it stops vacuoles from being able to store nutrients correctly, disrupting the energy supply to mitochondria. The mitochondria, in turn, degrade.

Calorie Restriction

By stopping the vacuoles from becoming less acidic via calorie restriction, the mitochondria did not degrade and the yeast cells live longer. Bottom line is that it was this storage function, not protein degradation, This appears to have caused mitochondrial dysfunction in aging yeast cell mitochondria by boosting the acidity of the vacuole.

Practical Application

Occasional light fasting may lengthen your lifespan and help you Thrive. However, as Duncan Crowe has pointed out: more broadly, fasting also stresses your entire body and most likely, your liver. So the fasting effects may cancel out each other out and stress causes many of our health problems in general.

Fasting

Generally, conventional wisdom is that a Friday fast or partial fast is fine for healthy adults. However, if you are on the AHP, you should be able to Thrive with no extremes. On the other hand, I do it without even thinking about it just because I get lost in my work... And when that happens, my love of what I am doing displaces my desire for food. It is difficult or impossible to measure the effects of light fasting, but the above story indicates that it is a good thing to do. Also, read Jon Barron's report on fasting. Jon is the reigning expert on fasting today, in my opinion, and this report explains the various types and their effects quite nicely: http://www.jonbarron.org/detox/water-fasting-and-juice-fasting?utm_source=iContact&utm_medium=email&utm_campaign=Jon%20Barron&utm_content=Biweekly+Newsletter+3%2F11%2F13

Too much of a good thing is a bad thing... Food in this case.

Now, I notice that Dr. Mercola is advocating missing breakfast... Just for the above reason... As another piece of conventional wisdom bites the dust. So I ask: Did Cave Men miss breakfast? Chances are that most missed plenty of meals. They were busy people for sure and they unknowingly benefited from it, but the thinking varies on this also.

12. LCD (Low Carb Diet)

Now onto the second of the three stars in the AHP, the Low Carb Diet. Notice that as soon as we move beyond the cellular level of discussion, the liver becomes the organ of focus:

The Scientific Basis

Much has been written about the Low Carb and Cave Man Diet and its relative, the Adkins diet. The Cave Man Diet is generally agreed to be based on the idea that ancient man ate in a way that his surroundings required. Certainly, grains have been in our diets for at least a few thousand years, but have we grown adapted to them to a great degree?. Since he was a hunter gatherer and never farmed, he never raised grains until probably about 2000BC and, more probably, seldom gathered them.
http://en.wikipedia.org/wiki/Paleolithic_diet

This idea has merit. However, while the LCD may have its roots here, our version, the LCD is really more scientifically based. It grew out of our earlier group discussion on the optimization of:

- Liver Balance and detoxification (as with GMO foods, etc)

- Glutathione Production
- Glycine Production
- Lowering the three Detoxification Pathways (our three stars in the AHP)

Oxidative Stress

All of these, as we have mentioned lead to Oxidative Stress. Oxidative Stress is the thousand-pound gorilla in the room. We theorize that the AHP minimizes this and, thus, is the key factor in Anti-Aging or staying young and healthy. With bells and whistles added, this requires generally <u>avoiding</u> all:

- Packaged Foods
- Refined Foods such as Table Salt and white sugar especially.
- Sugars and especially high fructose corn syrup, except as contained in whole Fruit.
- Too much High Sugar Fruit such as Bananas.
- Foods cooked higher than 140 Degree F whenever possible.
- Grains and Cereals.
- Vegetable Oils especially grain derived Oils
- Corn, Potatoes and Canned Vegetables (for many reasons).

Balance

By it's very name, the aim is to avoid carbohydrates/sugars whenever possible. The aim is to allow us to be able to take in Protein at a higher than normal rate, knowing that protein, itself, is an additional load on the liver. We balance the additional protein load with a very low carb/grain/sugar load. Therefore, the net effect is to minimize stress on the liver while increasing glutathione to glycine production, and maximizes body electrics, health, and, ultimately, longevity.

Sorting It Out

More specifically, the LCD, allows sugars and carbs bound in vegetables & fruits from organic soils, some nuts, meats and optimum fats. Yes, fats are actually required and these are not what are commonly referred to as "good fats." However, saturated fats are especially necessary to health. Keep in mind that all animals store toxins in their fat, so herein lies a secondary problem, but Coconut oil is a highly sought clean source as is Cacao Butter. Finally, we must provide enough live enzymes to keep things moving (more fully explained later). The idea is that you have to tailor the LCD to fit your body's needs. No one but you can make that final determination. Again, go slowly as explained earlier, Carefully tailor your LCD as with all parts of the AHP. Again, as stated earlier, when you make two major changes and one affects you negatively (or positively), It may take quite a while to sort out the cause even with just one change.

The Pottenger Cat Study

How important are enzymes? Read the results of a study done from 1932-42 on some 700 cats and how cooked meat decimated them and finally wiped them out in four generations http: //en.wikipedia.org/wiki/Francis_M._Pottenger,_Jr.

From Wikipedia:

- By the end of the first generation the cats started to develop degenerative diseases and became quite lazy.[citation needed]*By the end of the second generation, the cats had developed degenerative diseases by mid-life and started losing their coordination.
- By the end of the third generation the cats had developed degenerative diseases very early in life and some were born blind and weak and had a much shorter life span. Many of the third generation cats couldn't even produce offspring. There was an abundance of parasites and vermin while skin diseases and allergies increased from an incidence of five percent in normal cats to over 90 percent in the third generation of deficient cats. Kittens of the third generation did not survive six months. Bones became soft and pliable and the cats suffered from adverse personality changes. Males became docile while females became more aggressive. The cats suffered from most of the degenerative diseases encountered in human medicine and died out totally by the fourth generation.

Optimum Health

With the Low Carb Diet (LCD), our aim is optimum health, not weight loss. However, it

should not surprise anyone that implementing the LCD results in "optimum weight," since your body knows where it should be, even if your brain disagrees. So the LCD will prove to be the way out of obesity for you and, of course, the AHP is about optimizing everything in this process. Never piecemeal, the AHP is an overall approach to optimum health. For more depth on this read: the unparalleled nutrition book, "Deep Nutrition," by Dr. Catherine Shanahan, MD, giving more depth than a stack of nutrition books.

All Calories Are Not Alike

The below 2012 study From WebMD reinforces what we are saying. This, of course, flies in the face of conventional wisdom, since a calorie is a unit of heat, but it is true:

> "From a metabolic perspective our study suggests that all calories are not alike, Ludwig tells WebMD. "The quality of the calories going in is going to affect the number of calories going out." Therefore, Different Diets, Different Outcomes.
>
> The study included 21 young adults who originally lost 10% to 15% of their body weight on a diet that included 45% carbohydrates, 30% fats, and 25% protein.
>
> During the course of the study, the participants followed three different diets for a month each, including:
>
> - A low-fat diet, which included mostly whole grains, fruits, and vegetables, where 60% of daily calories came from carbohydrates, 20% from fats, and 20% from protein.
> - A low-glycemic-index diet, which included minimally processed grains, vegetables, legumes, and healthy fats, where 40% of calories came from carbohydrates, 40% from fat, and 20% from protein.
> - A very-low-carb diet, modeled after the Atkins plan, where 10% of calories came from carbohydrates, 60% from fats, and 30% from protein.
>
> In this study participants ate about 1,600 calories a day on each of the diets and the amount of calories burned was measured using state-of-the-art methods.
>
> The testing confirms that they burned about 300 calories more a day when following the very-low-carb eating plan compared to the low-fat plan, and about

> 150 calories more on the low-glycemic index diet compared to the low-fat plan.

Finally, as Dr. Shanahan points out in her above book, fat itself slows down calorie take up. So a cookie made with natural fats is 4x lower in available calories than an equal amount of pure liquid sugar calories!

Make It Fit

The LCD is actually a combination of the last two with modifications that best suite you. Remember, you are unique and, particularly, with diet, tailor what you eat to what makes you Thrive. For instance, if you never drank milk, it may be for a reason. Were you just listening?

The Dr. Terry Wahls TED Lecture

Many people in our discussion group refer to this Wahls video on natural food... This is Not the LCD but it is a telling story and worth the listen. Once you have absorbed all of the above: Watch this TED lecture (short) about a Dr. Terry Wahls who cured herself of MS using a natural food diet.

http://www.youtube.com/watch?v=KLjgBLwH3Wc

Notice how often Dr. Wahls references Sulfur to mitochondria? As discussed earlier, the mitochondria ***are*** the energy plants of the cells. They are both electrical and mechanical in function and are probably the key players that are responding to liver health. My premise here is that they also function at the Quantum Level.

13. It Really is the Liver

Finally, the Liver is the bottom line (but there is plenty more to this story from here):

The Liver is the most important and complex organ in the body, period. Relatively, the heart is just a simple pump, but there is much more to the heart and it works on a vortex (vibratory) flow that few understand. While both are critical to life, when the liver is stressed it throws off the factors that mainstream medicine looks at so carefully, like

HDL/LDL ratios (but there is only one kind of cholesterol as discussed elsewhere).

The point here is that when the liver is kept lean and detoxified, the heart and the entire cardiovascular system will respond with life-long rewards. So the suggestion is that we redirect our attention to this most vital organ and the others will fall into place. While this may sound like a simple solution, it is not, but it is far more direct and it actually will resolve most health problems first-hand... And that is what we re after with the AHP. We are simply not interested in band-aids under the AHP.

It's Also The Liver, Generally

Before we move into the more complex liver functions and health considerations, lets review some of the general functions of this incredible organ. It must

- Produces and processes enzymes.
- Metabolizes complex proteins into simple compounds (Recall that Whey is high in protein).
- Convert carbs...
- Perform a balancing act between the various dietary inputs.
- Store Fat: Simple carbohydrates are converted here into triglycerides, which are then stored in the liver as fat. The more fat stored, the harder it is for the liver to perform, so this is a function that we want to avoid, but these are mostly based in transfats today and much harder for our body to deal with.
- Process Drugs: All drugs and chemicals are processed, purified, and refined here for elimination. This is a function that can easily overload it also.
- Process Chemicals: Toxins, heavy metals, and pesticides... Everything we breathe, eat, or absorb through our skin is purified and refined by the liver.
- Process Alcohol: Just another Chemical that it is asked to process, but with grain based or fruit-based sugar... simple carbohydrates. So alcohol and sugar are doing double duty in overwhelming the liver. If you entirely overload it, cirrhosis (hardening) occurs; when it dies and you go with it (One beer is my limit, ever, for this reason and others).
- Process Vitamin Isolates. Many vitamins are seen as harmful chemicals to the liver and must be eliminated... Especially, artificially manufactured vitamins

contribute to liver overload. They cause worse effects, but this is true.

The Liver in a Nut Shell
The LCD reduces liver stress at the carb side, while allowing the Whey to provide the liver additional fuel on the protein side. Then, the MSM is allowed to do its own magic as described later in a very complex process. The claim then is: the LCD increase liver

function exponentially compared to anything ever previously proposed

More Second-Hand Proof: Jon Barron's report directly ties Liver health to Heart Disease and Type II Diabetes. In two related studies, Heart Disease was shown to be up three times (3X) and Diabetes two times (2X) above normal occurrence due to Liver overload from the plastic B.A. Jon is probably not aware of the AHP, but he gets it. When you hear what he says, adding what you know, you will completely understand why "It IS the Liver Stupid." Liver overload is the fundamental cause of most diseases, period. Hence the title. Listen to his podcast on Liver and Heart health at:
http://www.jonbarron.org/sites/default/files/talkr11.mp3

14. Undenatured Whey Protein Program (UWPP)

My friend Duncan marketed **Immunocal** and this form is superior to all others of whey, no question. I have never tried it, but that company provided the basic research that he used in his UWPP Protocol. This whey component is a pure, high quality protein source. It delivers sulfur and enzymes when it is carefully handled. Other sources are very good though and especially from New Zealand goats. (5th Edition)

Undenatured means that it has not been heated beyond 140deg F in the processing. There are many whey sources out there that have been prepared at high temperatures and thus have been denatured. These must be avoided entirely. Enzymes are very delicate, but sulfur is even more so. Frozen smoothies are good food. However, as discussed later, they not a good source of available sulfur, because virtually any processing helps to destroy the delicate sulfur.

Immunocal Website

website:http://www.immunotec.com/IRL/Public/en/CAN/item_0005000.wcp
://www.immunotec.com/IRL/Public/en/CAN/item_0005000.wcp

Forms of Whey

In the interest of economy, we will discuss Now Whey, which Duncan recommended to me.. Keep in mind that it is not necessarily the best Whey choice, but it works.

A 5 Lb. bags of Now Whey that I used would commonly last me about a month, but I used it at a higher rate than Duncan recommends. Today, its fresh, organic Goat Whey from Swanson (in one pond lots), but the Now brand worked just fine. (5th Edition)

Our state laws require our dairy cream separated by my good friend, mentioned below, at the Rehoboth Dairy and mix it per my recipes (see Ch. 43, Recipes). Also, I rely on MSM for my sulfur and I use lots of it, but enjoy the Whey for its protein benefits.

In my case, frozen whey/fruit/protein shakes (or smoothies) supplant what used to be Sealtest (as I recall) ice cream eaten at the Diary Bar in the old days of Junior High School, or Tastee Freeze (soy based) a few years later in High School. Then I moved up to Bryer's Ice Cream after graduation and college. Then, on to one of many great homemade Ice Cream stores around the Boston Architectural College. On moving here, I went to Kings Ice Cream down the street, then on to the Dairy Farm Ice Cream about a mile from here, all bad choices for their **sugar** content. (5th Edition)

So, for me, things have changed quite a bit in terms of food quality. For a time when I thought that Tastee Freeze was all the rage, I was scraping rock bottom, as I consumed what amounted to soy bean products (with little or even negative food value). However, all of the above ice cream contained plenty of sugar (and probably other products that were no better). Today, I freeze these whey/fruit/protein shake/smoothies into ice cream using "organic" Rehoboth Dairy Cream. Now that is getting somewhere. Now, what used to be a treat has become a staple and that keeps a permanent smile on my face. It's not as good as sailing a fast sloop in the B.V.I., but, given the health benefits, maybe I can sustain my ability to do that for quite a few more years. In any case, the bottom line is that this whey stuff is great tasting, especially, when you add Rehoboth Dairy Cream and it actually aids in anti-aging and health... what a deal!!!

15. The Anti-Aging and Health Program (AHP)

So with the above three stars, we have found the way to Anti-Aging and Health:

My premise here is that we can reform mitochondria and, possibly, lengthen Telomeres and thus extend lifespans exponentially with a combination of our basic concept... and the very foundation of this book is the formulae:

LCD + UWPP + *HL* MSM + Energy (raising your spiritual vibrations) = AHP.

AHP, the Anti-Aging & Health Program entails doing everything in this book at as high a level as possible. That is to say that when the four items are combined with the basic ideas presented herein, in a way that best suites you, you create this single program that will allow your body, mind, and beyond (discussed later) to reach optimum health and vitality or... Thrive!

Sulfur

Generally, Organic Sulfur opens Electronic Pathways and is a Cellular Oxidizer: So mind/ communication pathways open better with cells when more is available. As mentioned earlier, these electronic pathways function on the Quantum Level. If they become impaired in any way, this lack of communication becomes the overall problem.

By the way, even though whey contains sulfur. Sulfur is basically unavailable once a food is canned, refrigerated, frozen, heated or otherwise processed, so it is likely that whey ice cream and smoothies no longer deliver useful amounts of sulfur, sorry to say.

While MSM always delivers sulfur, no other dietary component is so easily lost. Any food grown in fertilized soil or sprayed with pesticides is usually devoid of Sulfur. So Sulfur has been nearly completely eliminated from the contemporary diet for everyone and this was initiated with the rise of modern farming and refrigeration. That process was started more than a hundred and fifty years ago and has steadily moved to where we are today. Knowing all of this, it is difficult to discern why people actually still exist at all, let alone live as long as they do on average. It is safe to say that anyone reading this for the first time is lacking, but the next component is equally key:

The Methyl Component

The real natural life span for a healthy human being, not long ago, was thought by most scientists to be completely limited to 12 x 12 = 144 years as limited by telomere length, but since, we have this **methyl** component (**MSM**) combined with **enzymes** and **sulfur** available and now understand more of this; this entire premise may be wrong as we kick the can down the road and I suggest that it is by a factor of ten or more (5th Edition)

The Grand Experiment

Think about it: Just maybe, the O.T. Christian Bible references were actually correct. In fact, their yearly age computations (a common explanation) were dead equal to ours, so sun cycles were the same as ours. If their food quality was superior (and we know that it was), and their water sources actually contained the altered qualities of the water explained in my latest book, "Pretty Fine Sex is Spiritual" and they did not overeat (they probably seldom could)... they were very healthy. What is more, the good Electro Magnetic Frequencies (EMF) produced by the sun may have changed just as has the skin color of those of us who left Africa. Dark skin pigmentation is more efficient and less destructive in absorption http://magnopro-usa.com/emf-good-vs-bad.htm. of good EMF even though it absorbs less Vitamin D (an adaptagen). Furthermore, in this scenario, the clothing necessary in northern climates is a decided detriment, as is a sun block and sun screen as discussed elsewhere here. So, given these factors, the lucky individuals who survived the pitfalls of tropical climates just may have lived a very long time... barring becoming dinner for lions, mamba bites, and other accidents. So eight hundred years could have been a rather normal lifespan (5th edition).

Much has already been written of mitochondria and telomere length,, but these two issues, worthy of more study, could change our understanding of aging completely. Certainly, even if these yearly computations do not or can no longer work out, our wellness is bound to them and this book will quite possibly tell you why (as I write this, I trim back my finger nails for the second time in a week... and not out of vanity). So please help me test this real possibility by joining this **AHP.**

"You" Know

This **AHP** is a complex story, but in practice, anything that is inherently true must be simple as long as you, the individual, generally stay with it. My point in saying "generally" is that if you eat one Godiva Chocolate this year, you are not going to join the dying crowd, but if you "get hooked" and eat a box each month, you will have joined the losers. Spiritual truths are not complex. We intuitively understand the truth when we hear it, assuming that our attitude is correct. That is, we are ready to listen. If you hear someone making it really difficult, they are among the lost... Run from them. It's a lie.

16. Why AHP Works (5th edition)

Now more on this AHP... the overall program and its nuts & bolts:

How Sulfur Is Lost

The most weak pathway in our health venture for most people alive today, from a dietary standpoint, is sulfation. Sulfation is responsible for the negative transformation of neurotransmitters, steroid hormones, drugs, industrial chemicals, phenolics (compounds derived from benzene (commonly used in plastics), disinfectants, and pharmaceuticals), and especially toxins from intestinal bacteria and the environment as noted above.

Organic sulfur from whey and MSM, then, as an adaptogen, serves as the primary detoxifiers is this process. That this, sulfur must come from our diets through food is a given, but it is especially available with the AHP. So we need more sulfur and we are getting less. Can you see where we are going wrong here (5th edition)?

If your exposure to substances that need to be detoxified via the sulfation pathway is high (always the case in developed countries), but your sulfate reserves are low due to an inadequate (SAD) diet, you will not be able to break down toxins. The results are inevitable, unless one of the diseases listed below gets you first: You will be rewarded with 50 or 60 years of good life and, maybe, another 20 or 30 of pain and suffering. But this is what we have come to accept as normal circumstances in developed countries and the U.S. just came in last on that list according to reports by Jon Barron and more recently by Dr. Mercola. Interestingly, since the first version of this book, this story has worsened measurably as we accelerate this downward spiral (5th edition)

.

The Detoxification Pathways

Sulfur

Without an adequate supply of organic sulfur via AHP, a variety of illnesses await, including: Alzheimer's disease, Cancer, Heart Disease, Parkinson's disease, Motor Neuron disease, Autism, Biliary Cirrhosis (Liver Disease), Osteoarthritis (OA), Rheumatoid Arthritis (RA), food sensitivities, chemical sensitivities and chronic respiratory conditions. Add these to a multitude of other chemical sensitivities and chronic conditions mentioned elsewhere herein and you realize that this AHP is not only important, but essential to any serious overall health and wellness program. Also, sulfur itself is probably the most important detoxifier, but below are several others:

Amino Acids

We manufacture five amino acids that form a third detoxification pathway: Taurine, Arginine, Ornithine, Glycine, and Glutamine. Of these, Glycine is the most important for the neutralization of toxins. In most cases, the body cannot make enough Glycine to keep up with its own detoxification needs. Though not considered an essential amino acid (because the body can make it), Glycine production depends on an adequate intake of dietary protein (Whey is a start). I supplement Taurine, Arginine, and Ornithine. Glycine & Glutamine are inherently part of the AHP and so it is recommended that we supplement the other three just to be safe. An excellent choice is Nutra Bio Arginine Proglutamate Plus Lysine, a growth hormone releaser, then add Ornithine separately. (5th edition)

Glycine Depletion

Glycine can be depleted by any number of lifestyle stresses. Especially the Benzoates, found in soft drinks. Bind them with Glycine and you rob the body's store of it. One study found that people who consumed a large number of soft drinks had problems breaking down toluene, a common industrial organic solvent. Knowing this, and knowing that soft drinks are primarily fructose injections, do you ever "need" another? Interestingly, our, so called, sports drinks are hardly more than modified soft drinks.

Aspirin and NSAIDs also slow down this detoxification pathway because they compete for available Glycine in the liver. So avoid them. White willow bark, which Bayer synthesized and patented is the organic basis for aspirin. The Romans used it and guess what? It still works. But easy on that too. Just because it is an herb does not make it good for you and it is not, particularly, but synthetic Bayer is far worse (5th edition).

Furthermore, all drugs have toxic components and as noted above must be filtered out by the liver. We hear lengthy commentary on this daily at the end of each drug advertisement. AHP will help your body deal with their effects, but it's nearly always far better to avoid over-the-counter drugs as well as prescription drugs with rare exceptions and, no, your government is not looking out for you in this regard.

Energy from Methylation (The Supercharger)

We have a new set of pathways that have gone unnoticed until a few years ago and is still not recognized by those who rave about MSM and DMSO. I now dedicate a Chapter 11 to these later in this book. This begins to explain the incredible results that people obtain from the addition of HL MSM.
It was first brought to my attention by a member of the Coconut Oil Group in about 2013.and further emphasized by my friend Dr Richard Price when he became familiar with the importance of and manipulation of mitochondria. Rich was/is using my developed fractals to do this, but methylation is the key to the three carbon cycles in the body and Sulfur is a key element to one of them. I further emphasize this later, but this how MSM brings its magic chemistry into play and this is the why we see this supercharger work.

Using AHP:

- There must be a noticeable improvement in the detoxification capabilities of your body.
- Overall health must improve to a degree, even if you keep some bad habits.
- Given any life threatening condition, you are guaranteeing your body that it will have a chance to survive and Thrive.
- The above-mentioned diseases put you in survival mode and AHP takes you out of it.
- As a biological organism, your body strives to stay well and generally it is has a high capacity to do just that. The more you help it with the AHP, the better chance it has of keeping you well and avoiding the diseased condition.

Bottom line: Given this AHP combination, all diseases become less of a challenge through the elimination of simple carbohydrates, feeding the liver, and resulting in lower oxidative stress.

Immusist

http://immusist.com/ is a surfactant (like soap, reduces surface tension) blended beverage concentrate and is another way to further enhance a compromised immune system in order to move it toward optimum health.

17. Enzymes, Anti-Oxidants & Liver Detoxification

Much has been written of Enzymes and Anti-Oxidants, but below we begin to see just how they function in terms of the liver... The place where we must make a difference if we wish to Thrive:

Enzymes

Basically, your body doesn't keep old molecules around for long. Even those of great value. Enzymes, are constantly being disassembled and reconstructed, recycled, or eliminated. Enzymes are slower to go than most, though, and we can build them up to a degree. I actually supplement with some thirty different digestive enzymes just to help lighten the load. You could not supplement with too many enzymes. They help the liver do its tricks to keep you vibrant and well.

The Liver Detoxification Process

Thanks to the detoxification process and enzymes, the liver is able to break up the most toxic and dangerous of molecules and evacuate them. Enzymes act as catalysts in this transformation process. We generally think of digestive enzymes (when we think of them at all), but there are literally thousands of different enzymes, each with a unique role. More on Enzymes later, but they are central to our existence and survival. Enzymes are certainly one of the key elements in health, anti-aging, and wellness and your liver can catalyze several million enzyme reactions per second, but rates depend on conditions. Of course, our aim is to optimize those with the AHP and we can do just that.

Think of this detoxification process as a two-phase wash cycle (very watered down, lol):

Wash Phase I:

Enzymes are the natural surfactants (again soaps) that reduce grease into droplets. They remove the impurities that the water alone can't remove. In the first part of the wash cycle, enzymes break toxins down into intermediates. Some toxins are ready for elimination at this stage, but others require a second detox cycle as below:

Wash Phase II

From here begins the more complex cycle. Here toxins are broken down further and detoxification becomes the main concern. In Phase II, these intermediate compounds are routed along one of six chemically driven detoxification pathways, where they are further broken down, and bound to specific types of' protein molecules which act as "escorts" guiding them out of the body through "Conjugation." Conjugation is the direct transfer of material by cell-to-cell contact. It is a bridge-like connection between two cells which allows them to exit toxic material through excretion. Interestingly, this transfer occurs between bacteria. Bacteria also are used to clean oil out of marine systems. They are our most important of natural environmental clean-up systems and have been given a bad rap by scientists in the past. You would simply die without them as discussed elsewhere herein.

Glutathione Detoxification

One of the most important systems in this Phase II which Duncan Crowe often called the Glutathione conjugation pathway. The key here is that Glutathione detoxifies our bodies of deadly industrial toxins such as PCBs and breaks down any number of carcinogens. This activity accounts for up to 60% of the toxins excreted in the bile. Glutathione, **the master antioxidant**, also circulates through the bloodstream combating free radicals. No other conjugating substance is as versatile as Glutathione and the body's supply of it, mainly produced by (you guessed it) the liver, is easily depleted... unless we enlist this AHP. Given this fact, guess what disease results when glutathione is depleted? See Wiki for more on this: **http://en.wikipedia.org/wiki/Glutathione_S-transferase**

Sulfur Absorption

This is essential to providing the above pathway. The UWPP alone, could serve as the only nutrient in this glutathione process, but, of course, *HL* MSM Protocol alone is 1/3 organic sulfur with 2/3 methyl goodies Without these two, exposure to high levels of toxins could quickly exhaust all reserves of Glutathione, increasing susceptibility to the above conditions. Once it gets a toe hold, Cancer (and its conventional treatments)... and all Chronic diseases including HIV and Cirrhosis quickly deplete Glutathione reserves. In fact, it could be argued that all diseases are the result of deficiencies of these vital components. This 2/3 methyl component is the magical part completely overlooked by today's proponents of MSM on the web, by the way. (5th edition)
http://www.healthy.net/scr/article.aspx?ID=2066
In animals and cattle: **http://agriking.com/uploads/2013/12/Advantage_Jan2014.pdf**

Oxidative Stress

A constant given, Oxidation occurs as a result of any number of things including many normally considered healthy... like excessive exercise.

On the other end of the spectrum: Just chillin' out just watching TV & drinking a beer (wouldn't you know it?) causes oxidative stress. In fact, we could never totally avoid oxidative stress in our lives. It is always occurring and is really just a matter of degree... and it always produces free radicals and stresses the liver. However, we can beat it with the AHP! Read more on Oxidative Stress at:
http://en.wikipedia.org/wiki/Oxidative_stress

Living Is Stressful Oxidative stress can either cut-back or totally block Glutathione production. Done long enough, it will deplete Glutathione in the blood. So as you can see, what is considered "normal living" is "life threatening." Hey, you could die from this! My premise here is that most people do, but it takes many forms down the road.

Reflecting, you have to wonder how our bodies have survived doing this balancing act with the low levels of nutrients that we consider "food." The AHP is the only healthy path to surviving this maelstrom of problems facing us on this planet, especially, since we are being given bad and contradictory advice by our own FDA and experts who pass on their advice. Interestingly though, this AHP entails eating and drinking some of the most tasty foods on earth, when used creatively (doing your job).

18. Drs. Seneff/ Mercola video/ the Organic Sulfur story

Dr. Seneff & Dr. Mercola have done several new interviews since this original. If you are interested in knowing what is really going on, watch them all. There is a wealth of information on these video and Seneff is a huge proponent:
http://www.youtube.com/watch?v=5QUChSlUEH0

(if this does not work for you, just go to You Tube and type in Dr. Seneff... please don't cheat, watch them all if you are serious about this. (There were seven parts, last check, and they have added several more since. Seneff is fascinating. Do not miss a single one if you can help it).

This is a long, but most interesting video. It should be watched, especially, by all over the age of 50, even if they have zero health problems. It could add years to your life for almost no cost especially when combined with this AHP. Dr. Seneff has learned quite a few things about sulfur and certainly has turned some new stones when it comes to Heart Health and Diseases. Furthermore, Seneff has come up with some very important dietary relationships between cholesterol and sulfur that make great sense and none conflict with the facts and recommendations presented herein.

Actually, I have followed her recommended diet quite closely (the LCD) for sometime now and feel better for it (long before I had heard of her). As kind of proof to her opinion, for those mainstream doctors: one little nasty fact statistically (and these are readily available), is that the higher your cholesterol levels are once you reach age 80, the longer that you will live *(actually, this age is being lowered as we speak. But, so will average lifespans for all who do not to embrace AHP, since these components apparently set the maximum age bar).* So now we begin to learn that high cholesterol is nearly always good.

Even though I have followed the results of organic sulfur for years, the mainstream medical community is just now learning the incredible value of it. I have been researching it on the web ever since I had my wonderful results with *HL* MSM years ago. Seneff's list of sulfur's values is just the beginning to take shape in earnest as we progress.

Recent animal studies indicate that sulfur is most likely a cancer shield, of sorts, as the above glycine story tells us. It helps make arteries more pliable. It helps repair any number of diseases related to improper tissue replication and much more. Seneff gets her information first-hand and that is admirable, but much of what she has discovered is repetitious as would be expected since we are reporting ... so we repeat ourselves.

Seneff Talking Points

Her interesting talking points worth considering are still often just opinions, but they become much less so when you read it all in one place. Things begin to just jump out at you:

- Our joints are collagen which are high in organic sulfur. In order for our overall body to exist, it must "steal" sulfur from the joints. So feeding your joints (keeping an adequate supply) makes you healthy overall ***(I reverse this, herein, but the facts remain).***
- Soft water causes your body to loose sulfur, so if you have a water softener, you really need more (more on water generally in "Pretty Fine Sex is Spiritual."
- Eliminating carbs helps raise sulfur levels ***(allows you to process the sugars bound in fruits and carbs bound in vegetable better. I explain why elsewhere).***
- Eat highly saturated fat foods like coconut oils (and no vegetable oils, especially hydrogenated varieties)... and avoid processed sugar totally if your goal is health ***(you can take this further and avoid banana's for their excess sugar.***
- High sulfur and low carbs encourage weight optimization ***(carbs always become sugar)***
- Eating animal fat and red meats high in fat does not make you gain weight (discussed earlier herein, but keep in mind that poisons are stored in fats).
- Note that animals highly regard fat (and organs) for what it is and eat that first. It feeds your brain and keeps you well **(as discussed elsewhere).**
- Fried foods and corn oil, on the other hand, do make you fat, as does sugar, and especially anything containing high amounts of refined sugar or high fructose corn syrup (like sodas). **(now, add the GMO component of corn and corn oil).**
- Modern farming practices have decreased the sulfur content of our crops ***(and especially minerals, but she does not mention this, read Dr. Joel Wallach's site for that part***).
- Vitamin D & sulfur work together ***(Sunlight is the best source of D.***
- Aluminum Sulfate has been increased in vaccines along with mercury and both are potentially huge problems Older people should never get flu shots if this is correct.
- Autistic children always have low organic sulfur levels (think about this).
- Among its long list of virtues, Sulfur protects you from skin cancer ***(as does D3)***.
- ADHD and Alzheimer's may be caused by a sulfur deficiency and, as she says, they could actually be the same thing! ***(add depleted saturated fats per Wallach).***
- Antiperspirants actually cause a reduction in sulfur and a corresponding increase in aluminum ***(hmm... does this set off alarms?).***

Some opinions beyond Seneff's points
Seneff admits she does not know anything about MSM (other than it exists) and that it may heal organic sulfur deficiency problems. On hearing her interview and realizing the impact of organic sulfur on our bodies as she describes, it is clear that our AHP will positively affect virtually any condition.
This Mercola video repeats much of what Duncan has been telling us for years about oxidative stress and Glutathione conjugation, so it's not new. Add glycation and methylation mentioned herein to get more on that story.

Still, taken as a whole, Seneff's opinions are likely the best "somewhat mainstream" source of organic sulfur information (with results) currently available. Her opinions and first-hand findings drive home Duncan's with regard to liver health and balance.

Further Key Points Observations based on Seneff's interview:

- MSM is the way to increase sulfur level quickly.
- Chondroitin Sulfate, with its large molecules, is not really bioavailable and is useless.
- Glucosamine Sulfate causes gastric distress... so is ineffective in this manner.
- Eggs and all other foods are actually a poor source of sulfur compared to *HL* MSM.
- At least most of us, simply can't take sulfur in any form other than from plants, UWPP, and *HL* MSM. Excess sulfur rarely occurs, but is quickly disposed of by the body.
- I have been on *HL* MSM at ½ cup/ day for months on end with nothing but great results. As long as your stomach can tolerate MSM, and with most of us, it is not a problem, your results should equal mine. As Seneff says, we are all deficient in organic sulfur due to farming practices... Add commercial practices and food handling practices.
- As Seneff observes, sulfur apparently softens your arteries. I had discovered that indirectly because when people take MSM, at my recommendation, for their joints. They report that their blood pressure tends to decrease... a lot sometimes! The

body self regulates when the liver is properly nourished (again, it really is the liver stupid).

- From this Video, the AHP will actually CURE Alzheimer's, Heart Disease, Autism, and MS... and quickly (add saturated fat and organ meat).
- In addition, Cachexia (Or muscle wasting), now a serious problem for anyone over 60 even in less developed countries, should be stopped dead by the AHP.
- Whey on a daily basis is a great way of maintaining healthy sulfur levels and Dr. Mercola is correct on this... But isolates are a fine source of whey and Mercola sells concentrates.

The bottom line: Everyone, even kids, should be supplementing with at least moderate amounts of MSM and eating their "Curds &Whey." These are actually a cheaper and more dense organic sulfur source than the foods that Seneff mentions. We now know that the older population, because it is sick and dying, has generally proven that common food sources cannot deliver maintenance doses much less therapeutic sulfur doses.

19. It's a Mineral Deficiency

Below is a commentary on the "mineral king" and his opinions:

Veterinary Medicine: It is interesting to note that veterinary medicine is possibly the closest to any form of pure medicine left in the western world (except that pet foods are even worse than ours and they often sell and endorse them), since the drug companies have taken over most of what was once empirical medicine. That is, they do not have to answer to the legal issues of not being allowed to do what they have learned to be true through general practice as opposed to MD's who are now mostly drug salesmen.

Dr. Joel Wallach learned from practicing on animals (not even pets), where he could do things that would be frowned on by, "accepted practice" and could not be sued for doing what worked. He and people like him, carry the real badge of "Do No Harm." Interestingly, as a kid, veterinarians actually treated me on Grandpa's melon farm. I thought it was really a second rate deal then, but not so, now, as I look back. Hey, their stuff worked, but I knew that I was not a cow.

The main theory proposed by Dr. Joel Wallach, the veterinarian/ humorist, best known for the tape, "Dead Doctors Don't Lie" (it is still available online & fun to listen to) **http://www.majesticearth-minerals.com/** is that virtually all health problems can be traced back to mineral deficiencies and vitamins simply exist to aid in the processing of minerals. Today, we can add, based on the work of Dr. Dan Nelson, Dr. Rich Price and others who have shown by their work that minerals are, at the mitochondrial level, simply vibratory frequencies that never cross the microcellular barrier. That is, you never digest minerals per se. The mineral simply transfers its unique vibrations across the cell membrane raising the vibration rates of your entire body.

Taking the above to its logical conclusion, Dr. Wallach discovered in animal testing that true nutrition, then, is only about getting the correct mineral vibrations to the mitochondria where, basically, our energy is sourced. So the complicated (and energy intensive) digestive and elimination processes, in total, are only about this transformation process and vitamins aid in this transfer.

Furthermore, Joel points out that as long as you keep his point above in mind and: That you eat primary, organic foods, that is, you will never be concerned with such drivel as saturated fat content, your body will work to keep you well... a point repeated more commonly of late in the wellness community and herein. Having read several of his books and occasionally listened to his weekly radio show over the years, I laugh, but, also, generally agree with his theories. The man is a genius of sorts in that he keeps it simple and is always looking at base cause. As he says, farmers pay a lot of money for their bulls and they do not want to take any chances. We can all take a lesson from Joel in hearing how he approaches illness. His first question is always, "What mineral is lacking in this animal's diet that is causing the root problem?" Next, he then moves on to more complex possibilities, but his attention is always on, "Why is this animal not getting that essential mineral. Why is this animal's resistance so low that it is affected by that parasite, etc.? The key here is that when your vibratory rate drops, you are subject to disease.

Since first hearing him talk over twenty years ago, I have made it a point to take a liquid colloidal plant-based mineral source daily that contains the approximate 72 minerals that he proposes as essential to good health. No doubt, at least one major disease could probably be found that corresponds to each trace mineral... Save lead, mercury and a few others whose vibrations are poisons at even trace levels. .

To the above colloidal trace minerals that I take, I now add Selenium, as an antioxidant and anti-cancer, plus organic Boron, Iodine, Magnesium (especially, and discussed in depth below), and, occasionally, Strontium for bone health, as a result of our discussions in the Coconut Oil Group. One that I never supplement is calcium for reasons mentioned earlier except through moringa. You also should consider: trace amounts of Vanadium, for bone health and diabetes, Chromium, especially if you have diabetes issues, Silica/ Silicon which helps build collagen, Potassium, if you have blood pressure issues, and consider Zinc supplementation, during the cold season.

Absorption Rates You could add sea salt to the above mix. Sea salt contains trace amounts of these same 72 trace minerals that he recommends plus others. Wallach talks about how important salt licks are to cattle health and questions why people are not encouraged to use it. He then moves on to "absorption rates," and liquid colloidal minerals. He maintains that raw vegetables deliver these best. Joel also notes that vegetables must be raised in mineral rich soils because plants can never "make" minerals on their own, all simple wisdom. Stephanie Seneff alludes to this in her interviews, also. We can feed hydroponic plants raw minerals and our high energy water and they will take them up into their leaves for us to eat... as perfect, natural chelators (and vibrations).

Bottom line: Minerals are the balancing frequencies. They never cross the microcellular wall where our power plants exist. So per Wallach's conclusion, vitamins are only required as a delivery system for minerals thus are frequency enhancements to minerals. Herein, we discuss how these balancing mechanisms work as nature delivers them.

Vitamin D One exception to the above, which Joel's does not appear to address is that of Vitamin D. Most animals manufacture D internally, but humans do not. Therefore, D actually performs more of a hormonal function. However, recent findings on the myriad of D's effects indicate that it is essentially something apart from all others, a hormone/vitamin and a totally different breed of wellness essential. Whatever it is, I recommend that you pay attention to it and follow the Vitamin D Council online.

Alzheimer's Disease

One interesting side theory proposed by Dr. Joel Wallach is that Alzheimer's Disease is medically created and caused by our implementation of the low saturated fat diet (low cholesterol in blood, read statins, and low fat diets). He claims to have proven this years ago. Furthermore, he claims that he can manifest the same symptoms in some animals by depriving them of cholesterol/ saturated fats. You hear this more often now from others, but his original argument was that our brains are over 90% cholesterol and that the current low-fat mainstream diet deprives us of this essential nutrient, thus actually causing this terrible problem. Today, all of this is proving out!

So cholesterol is Blood, Liver and Brain food. It's amazing how ignorant we can be under the guise of intelligent science. Joel tells his stories with humor, but he is dead serious and it all makes sense.
.**http://articles.mercola.com/sites/articles/archive/2012/02/11/dr-stephanie-seneff-interview-on-statins.aspx** Statins, by decreasing fat, increase Alzheimer's disease problems. So a second opinion here that backs Joel's (and mine today).

Playing Wallach's card:

My son grew up with an ADHD problem of sorts. From age 2, he was a ninja fighter. He scared the daylights out of kids his own age. By the time he reached 3rd grade, he was the terror of his class.

Finally, I was forced by his teachers to have him visit a psychologist on a weekly basis or, I was told, he would be thrown out of public school. About that time, I started taking colloidal minerals and introduced them to him. He immediately got better and the teachers were thrilled, but it took me a month or more to figure out what had happened.

Eventually, my son forgot why he was using the colloidal minerals and stopped them. I have two fist holes in my walls recalling that era. However, I reminded him again and, today, he is back on them and is a wonderfully normal person.

All of this makes me wonder, could we substitute a trace or rare earth mineral for each of the drugs: Prosac Xanax, Ambien, Lexapro, Ambien, Benadryl, Temazepam, Mirtazapine and Trazodone? Probably, but then, where would the money go and what would psychiatrists do for a living without all these mentally disturbed people? It only takes one true solution to see that most are a scam. When will we see the facts?

Wallach's Videos and Books
Use (sea) salt. We must have it for stomach acid.
http://www.youtube.com/watch?v=_9_0gRpt_ok&NR=1&feature=endscreen
Buy his books and tapes
https://www.amazon.com/Dead-Doctors-Dont-Joel-Wallach/dp/0974858102
Dead Doctors Don't Lie:
http://www.youtube.com/watch?v=2-ltxZvVAMI
:

Relative to Wallach's work: It is interesting that a salt water fish will actually die if you do not add an amount of real sea water to its tank water. Does that mean that we, organisms that rose from this same sea, also need a certain amount of sea water (its not just salt water) to thrive over a long period of time? Since sea water is pretty inexpensive to come by, why test that one? I use it transdermally for that reason on a daily basis. Also, sea water hot tubs and swimming pools make great sense in this regard.

20. Magnesium- (5th edition)

This nutritional mineral for general health and particularly for heart health is generally deficient. Few people in our population come near getting enough of what we need to stay well. Herein, I give you the option that I find the easiest source for it on planet earth, but first lets discuss the available alternative chelates that I reported on here in my 1st edition.

Magnesium and Mineral Chelates

When you combine our low levels of cholesterol (Vitamin D) and lack of magnesium, you find our recipe for heart disease. There are many studies and indications that this is the heart mineral that most of us lack. In 1900, we absorbed an average of ½ gm of magnesium daily, but that number has dropped to less than 1/4 gm because the magnesium with sulfur, in soil has been depleted over time. Now add the fact that we are encouraged to avoid sun and wear sunscreen. This is the killer combination

As with many minerals, modern farming practices depleat and never add minerals back. Still, heart attacks, while they are far more common today, still occurred when people lived and farmed the same soil year after year. So this is another case of an insurance policy that you cannot afford to not take out on your body. At a minimum a gram a day, but the problem is getting that level into our systems easily through supplements. But read on my trick that it is easy and tasty, Cacao.

CFS/ Magnesium: An underlying magnesium deficiency, even if very mild, can result in chronic fatigue and the symptoms known as Chronic Fatigue Syndrome (CFS). In addition, low red blood cell magnesium levels, a more accurate measure of magnesium status than routine blood analysis (which are poor measures since the body relocates it according to need), have been found in many patients with CFS. The literature demonstrates that magnesium deficiency is not necessarily due to low dietary intake, and several studies have shown good results with supplementation improving magnesium stores.

On Magnesium Chelates: With oral magnesium all forms become laxatives at some point. Experts argue about which form works best... Here is an example, but if you have problems with chelates, skip down to my current choice, again Cacao.

:

“Studies indicate that magnesium is easily absorbed orally when it is bound to aspartate or citrate (chelated). In addition, both of these compounds may also help fight fatigue. Aspartate feeds into the Krebs cycle, the final common pathway to chemical energy or conversion of glucose, fatty acids, and amino acids, while citrate is itself a component of the Krebs cycle. Krebs cycle components (including aspartate, citrate, fumarate, malate, and succinate) usually provide a better mineral chelate; evidence suggests that these chelates are far better absorbed, used, and tolerated compared with inorganic or relatively insoluble mineral salts (such as magnesium chloride, oxide, or carbonate) (From the makers of Opti-Mag Aspartate... but fairly accurate).

Mineral Chelates:

Actually, one could consider plant based minerals as natural chelates, but we are discussing chemically bonded, man-made chelates here: Chelates may be correct for some but generally wrong for others. Magnesium Chelates have not proven to work well for me. I have not tried the various forms for a number of years, but plant-based minerals always have performed. The AHP actually makes your body more stable and more able to handle manufactured systems, which no one is telling you are natural.

Dr. Wallach seems to like them OK, but they have caused me gastric distress and I find totally plant-based forms to be the answer as per the Cacao entry below. At this point, I might go back and retry some chelated forms just so I know if my ability to take them has improved. But at least, maybe ten years ago, when taking them only orally, they quickly becomes too much of a good thing. I have found that two grams a day of natural magnesium, orally, is about all that I can ingest without problems, but Cacao has no downside and it tastes great:

Cacao - a Natural Primary Magnesium Source

So for the bottom line: Cacao is a farmed tree on tropical plantations worldwide, Theobroma Cacao, that grows to about 25 feet in height. The cacao seeds are commonly used to produce commercial chocolate. It has broad 3 to 8" leaves and produces small pink flowers in clusters that grow into orange to yellow seed pods.

The pod pulp can be made into a delicious and healthy juice. This has not caught on in the US, but it is prized in many tropical regions of the world today. The magnesium content if the cacao bean, juice and cacao butter varies, of course, but it is extremely high, and probably the best source in all of the plant world. Also, the butter contains 40-50% fat.

Known as theobromine, it is slightly water soluble, so is a very healthy fat source as well as a great magnesium source for humans.

We are told that chocolate (from cacao) is good for us by the mainstream. This is a trap for people with sweet teeth who use it as an excuse to consume sweet chocolate just as we do to consume alcoholic beverages. If you confined your consumption to unsweetened dark chocolate, you would be correct, but try that and, predictably, you will quickly find a different source of cacao. This stuff is nasty and no one does well with it, but read on.

Cocaine is a vascular constricting drug, first synthesized in 1860, from cacao leaves with the formulae C17H27NO4. It has some application in medicine, but is mainly used as a popular party drug. It to be avoided at all cost. It is metabolized primarily in the liver and produces cacaoethylene when taken with alcohol. This undesirably inhibits neurotransporters in the brain and creates a sense of euphoria. Again, avoid it as it impedes liver function and inhibits neurofunction.

Back to the positive effects of cacao: Theobromine also helps in the digestion of caffeine in the liver and, all the while, it reduces inflammation and promotes glycogen synthesis. While theobromine is weaker than caffeine, its effects are far better tolerated in quantity and it is helpful in treating bronchial infections and asthma, plus it just makes you feel good without the downside of caffeine. I find the pleasant (and legal) effects to last about four hours on average as it helps keep you well. Mix it with whey and moringa leaf, another tropical tree. I know just when my four hours are up after a number of years doing this, LOL. This is really cool stuff for sure. However, cacao is toxic to dogs and cats even in small quantities.

So how to use it? First, the ground beans (or seeds) when ground from the store like coffee, lose their component qualities fairly quickly (like coffee) unless kept in an airtight jar, so the best source of the beans is gained from buying the bean parts whole and grinding them in a coffee grinder into a small fine powered in smaller amounts. I actually like having a few larger pieces in the mix too, but suite your taste. They are crunchy like nuts and taste good in just the bean form, kind of fun to munch on occasionally.

Finally, the cacao butter is highly rated for cooking and obviously it is a wholesome food, even though it is classed as an oil. It melts at 32deg C and will withstand very high temperatures like coconut oil, so it is an equally great cooking oil. It costs about the same as the seeds. Also, it is a great base component for cosmetics and is often used in them. Look for it in your cosmetics if you are picky (as I hope that I can teach you to be herein).

The butter is not commonly rated as a nutritional food since it is an oil. But read this for ideas: **http://www.eatwell101.com/cocoa-butter-recipe-cocoa-butter-benefits**

21. Transdermal Magnesium

The Transdermal Magnesium System... that Dr. Mark Sircus, OMD, became famous for **http://www.amazon.com/Transdermal-Magnesium-Therapy-Mark-Sircus/DP/0978799119**
is known by his book title, "Transdermal Magnesium Therapy." The idea here is that your skin is a great delivery system. It seems to work well. The method that I have adapted from this in obtaining them is to fill up plastic jugs of Maine Sea Water and add Zechstein Salts
http://www.ancient-minerals.com/products/?utm_source=adcenter&utm_medium=cpc&utm_term=dr%20mark%20sircus&utm_campaign=AM%2BSearch%2BCompetitive%2B-%2BUS
plus Boric Acid to the boiling brine. Interestingly, when you boil the brine and add the salts, they do not separate out to any degree when they cool. I am not sure why this is true, but it is.

Application is easy*:* After a shower when your skin pores are open, just spray in on and rub it in. Soft skin areas are best, but all are good. Don't towel off; rub it in; it evaporates quickly and you feel dry, right away. It's a very nice way to start the day and you probably have just gotten about all of the Magnesium possible unless you happen to be a sea animal, you can take a swim in the ocean, or you have a sea water hot tub.
Does it really do what Mark claims? I am not getting muscle cramps and it always feels great... Plus, it too is inexpensive and simple. This is my measure of knowing if I am

getting adequate magnesium. It's quite simple. If you are getting leg cramps, you must find a better delivery system... one (or two in my case) that is effective for you. By the way, low potassium levels can also create cramps, but this is less an issue.

Calcium Supplements

From Consumer Labs, we get the following warning: "Calcium Risk for Women - A long-term study found that women with high calcium intakes from their diets were 40% more likely to die during the study than women with moderate calcium intake. Worse, the risk of death was **157%** higher for women with high dietary calcium intake who also took a calcium supplement." The bottom line for everyone is that you should avoid supplementing them as reported elsewhere herein. Get your calcium from natural sources and avoid such commercial products as orange juices (generally a bad idea in any case) that contain them.

22. Government Policy, GMO's, and Medicine

Now for some politics. These viewpoints are necessary for us to firmly grasp what is occurring in the real world relative to health & wellness. Without them., we can't make sound decisions relative to our health, and this misinformation is bombarded on us daily:

Politicians are fed with money from large drug corporations and, thus, the FDA is in bed with them. So, like it or not, fascism is alive and well. We simply cannot rely on what we are told by the FDA under any guise. For more on this, read http://michaelpollan.com/interviews/michael-pollan-debunks-food-myths/

Furthermore, as mentioned previously, scientific studies can no longer be relied on as "science." Some examples are represented here in The Lipid Hypothesis http://www.gnolls.org/1086/the-lipid-hypothesis-has-officially-failed-part-1-of-many/

GMO's & Drugs

- Why should a drug company be allowed any participation in any study that directly affects the public's use of that drug? **http://www.wellsphere.com/healthy-eating-article/the-demonization-of-saturated-fat/1218573**
- Advertisement of Drugs: GMO Foods (genetically modified foods): GMOs are repeatedly being uncovered by studies and observed and understood as a general health problem by responsible health care professionals. Still they are deemed safe by the FDA and other government agencies, while they are illegal in Europe and many foreign countries for sound reasons.
- The 2/13/17 issue of Consumer Reports argues the above problem coherently. We know that these products would simply disappear and the perpetrators, like DuPont and Cargill would, likely, vanish (or change product lines) if they were clearly understood by the public.
- The U.S. is a hold out on this issue... Just way too much has been invested at this point to consider the health problems that are bound to occur in coming years.
- Factory Farming: These practices sould change dramatically, particularly in their use of antibiotics, if the public knew the truth about Cargill, CAFOs, and Cows... the three C's, commonly passed on to us as normal "Farming Practices."

Of all of the things that effect our health, the one that we should have the most knowledge of is the food that we eat. Drugs are an extreme modification, generally of medicinal plants (but later on that). But currently, because of commercial interests, this is probably the area where we are most lied to and the FDA smiles on their every move. We may not be able to disband the FDA, but we must understand what is occurring and how to avoid the pitfalls of what they are allowing to occur if we are going to Thrive.

23. Astaxanthin: Organic Sunlight

http://www.livestrong.com/article/448563-astaxanthin-facts/

Seneff mentions Astaxanthin briefly in her interview. It is most often seen as the red pigment in sea run salmon and trout, but is derived commercially from algae biomass. Here is a story that illustrates its effect, first-hand:

In 2011, I spent an entire day at the U.S. Open at Flushing Meadows roasting in the sun with absolutely no protection and got way too much. So below is the happy outcome:

My nose was burned deeply as well as my upper arms above the tennis shirt line, but there was no peeling at all as a result of taking it for just two weeks. The red pigment, Astaxanthin, can be taken both transdermally and orally. In taking it, your system should have plenty of sulfur and Vit D3. In any event, this above experience adequately proves that the three together protect you from too much sun and the combination could protect you from skin cancer. I did not need anything else, no lotion and no hat. This is not a recommendation to overdo sun exposure, but it is a powerful recommendation for the three. A friend got skin cancer on his ear from the same exposure that very day and his wife now has a bench dedicated to her at our tennis courts (from it?).
Dr. David Williams first alerted me to Astaxanthin in the early '90s, but it was first isolated in 1975 by Professor Basil Weedon's group who proved its structure by synthesis. Http://www.drdavidwilliams.com/?key=201269&utm_medium=cpc&utm_source=paid_msn&utm_campaign=Williams%20Brand%20-%20Core%20Terms&utm_content=Dr.%20David%20Williams%20-%20Core%20-%20Phrase#axzz2Hgwv5xcI And why not? It is, quite simply, one of the most awesome supplements available and it does all sorts of great things in your body. Below is my simplified bullet list:

Astaxanthin's Benefits:

- Reduces oxidative stress . . . So **Liver Health** is key.
- Increases strength and endurance (2 - 8 times in human clinical studies).
- Helps athletes recover faster from workouts.
- Alleviates symptoms in patients with H. pylori (pre-ulcer indigestion).
- Protects mitochondrial membranes from oxidative damage.
- Boosts the immune system, increasing the number of antibody-producing cells.
- Prevents cancer in the tongue, oral cavity, large bowel, bladder, uterus, and breast.
- Decreases plaque formation, thus reducing risk of cardiovascular disease.
- May assist in the neurodegenerative diseases: AMD, Alzheimer's, Parkinson's, and ALS.
- Protects eyes and skin from UV damage, while protecting your skin from burning.

Because of these extraordinary characteristics, it is included in my own personal transdermal body cream preparation. Some reference links: http://en.wikipedia.org/wiki/Astaxanthin Visual Acuity and muscle improvement. What else it does it go is really unknown and when used transdermally, who knows? For certain, it helps aging eyes a lot and this is not often discussed by anyone. Find out and report back to me please.

24. Skin Wrinkling Tips Off Bone Loss

A study on Bone Loss plus some nutrient references.

Byron Richards, CCN reports this study on 6/29/11*:*
http://www.wellnessresources.com/health/articles/degree_of_skin_wrinkling_tips_off_bone_loss/ Pull up his entire article for his explanation, but below is a summary and conclusion. This, is indeed all true, but all of these are aging factors and they all stem from the same ones. HL MSM helps minimize all of these aging factors:

Study Indicates: The strength and growth rate of your fingernails and facial skin wrinkles helps predict the degree of bone loss and bone density accurately in women in their late 40s and early 50s. The study evaluated 114 women in their late 40s and early 50s who had their last menstrual cycle within three years – a potential time of high bone loss due to the sudden drop in estrogen. (Does estrogen really drop when you are on the AHP?)

Observations: More facial and neck wrinkles predicts lowest bone density. Also, wrinkling was predictive of bone loss at these measured sites: hip, lumbar spine and heels, and was independent of age, body composition or other factors known to influence bone density.

Why Collagen is Important: Bones and skin are comprised of 50% collagen. The main difference is that bones are highly mineralized. Thus, the same nutrient raw materials that support collagen also support both bone and skin formation. Also, the same type of free radical and other stressors that break down collagen do so in both your skin and bones. While there are differences between your skin and bones, they have enough in common to indicate that excess wear and tear in your skin may predict bone problems as well.

Collagen Carrying Nutrients There are many nutrients that are used for bone and joint health, including tissue-forming nutrients such as Hyaluronic Acid, Glucosamine, Chondroitin, Egg Shell Membrane. However, you simply can't get enough tissue-forming nutrients from these without severe gastric upset. Dietary protein and sulfur are low level examples of the AHP. The AHP provides skin, hair, and bone health via collagen synthesis and tissue matrix formation without this negative.

Skin Repair Factors: Other nutrients, in addition the above, that help include: Astaxanthin, Vitamin D3, Resveratrol, Vitamin C, Hemp and Grape Seed Extracts, Lycopene, Tocotrienols, and Green Tea. There is a more powerful and patented form of Resveratrol, Longevinex, resveratrol that includes a DNA repair factor as outlined below: **http://longevinexadvantage.com/** Is it better? Its hard to say, but it works well.

25. Reports by Others: Expert Opinion on MSM

Further discussion by experts on MSM and DMSO (edited):

Michael Harrah (in our group) sends us a good, but conservative view, of MSM apparently originating from Joshua Rodriguez Sosa (who has read this book):

- Methyl Sulfonyl Methane, better known as MSM, is biologically compatible with human sulfur which MSM gives up and is assimilated easily.
- Naturally detoxifying. It increases the permeability of the cell, allowing the water and nutrients to flow freely in cells while expelling the body's waste and toxins.
- Anti-inflammatory, relieving inflammation in rheumatoid arthritis and osteoarthritis while providing the critical sulfur necessary to regenerate the connective tissue and thus facilitating mobility and functioning of joints.

Connective Tissue serves to support to cartilage, bones, tendons and muscles. Therefore, MSM supports the functioning of joints, which allows for a proper body movement. Better still is that the MSM is compatible with all other nutrients that combat these diseases.

Combinations Help Thanks to these properties, the consumption of MSM is also recommended to assist in the recovery of sports injuries. In this sense, combining the MSM with other substances such as Chondroitin and Glucosamine provides excellent results and is a comprehensive treatment that appears effective.

(MSM has also been proposed as a treatment for interstitial cystitis (inflammation in the bladder wall that causes frequent and painful urination) as well as to improve mental ability and to promote the production of insulin. In addition, this supplement assists hair and nail growth and in general, strengthening the immune system. Some recommend its consumption to avoid the emergence of environmental allergies and migraines, and help control the stomach acidity. High Levels always help.

The importance of HL MSM supplementation is due to the fact that the sulphur content is unreliable in everyday foods. Also, the degree of assimilation can vary between different sources depending on their characteristics, in most circumstances today, they cannot meet the requirements of the body. In addition, the amount of sulphur will decrease during food processing.

L.c. Nut. Joshua Rodriguez Sosa
International masters' degree in nutrition and dietetics

Reciba un saludo afectuoso
de su amigo
L.A.E. Humberto Rojas B.
52-33-3685-0463
52-333-8150-542
HROJAS1206@YAHOO.COM

26. MSM/ Organic Sulfur 2015 (4th edition)

The Solutions Revolution - Patrick McGean - Clive De Carle
Watch the video: **https://www.youtube.com/watch?v=cg7Ubyt6O8Q**

Patrick McGean started with what he calls organic sulfur about five years after the author, but he has gained some creative insight in his venture. Is he totally correct? No. McGean does not recognize the Methylation Cycles in this contribution, but he mostly sees the key results from the use of MSM. His video goes a long way toward teaching you the benefits of what two tspn/ day of MSM will do. However, his recommendation will not likely help you much if you have severe health issues or are over fifty. Make that two tablespoons per day and you will venture into ultimate health quickly if you combine the cofactors herein. MSM clears the way for cofactors to play in by clearing the system of toxins and adding in high levels of sulfur.

His healing claims: Autism and how MSM lowers brain toxins; Tumors disappear; Lymphatic Pathways open up; Heavy Metals plaque disappears; Antiaging; Eye Health & repair; Cataracts disappear (DMSO drops are superior but more on that later here); Vascular health and regeneration and their fine capillary networks); Chemical Fertilizers: Inorganic C (ascorbic acid) is a problem.

Opinion: Based on my own experience as well as reader feedback and also on the chemical basis of MSM and DMSO McGean's claims are real. Also, so are those of the few distributors of MSM approved for human consumption. One product is approved by the FDA, thus their product cost may be ten times more including Patrick's. The FDA approved distributors do expensive studies to prove their product. Their claim is that this crystaline based, distilled MSM is superior.

Keep in mind that the chemistry is the same for all brands and it all works well from all feedback and my experience. Their further claim is thus that theirs is more pure, but this is unproven. Note that all MSM sold is essentially 99% pure (read the label). MSM is derived from DMSO extracted from wood pulp either by distillation or a water reduction process. They both have drawbacks, but the distillation can be more pure as it leaves no residue. I buy mine based on price and availability.

All MSM starts very clean, but obviously you want a product that has not been exposed to contaminants as MSM will grab them. This is one of MSM's most powerful qualities. Therefore, I buy a quick moving product in ten pound buckets and leave it sealed till I

move it into airtight glass containers.
So, bottom line, Neither the MSM that Patrick sells is better nor is Stanley Jacob's (that he mentions) better, however, both cost many times more than, for instance, AniMed for horses. Also, intake requirements change with age and conditions .To say that a teaspoon or two is wrong for a twenty-five year old in optimum health is incorrect, but its more incorrect to tell a person with terminal cancer to even just take a tablespoon or two per day.

Methylation

Methylation as discussed earlier, as it turns out, is the overlooked quality that all of the experts have missed with the possible exception of the late Dr. David Gregg reviewed below in Ch 27 & 28. However, even he was likely unaware of the Methylation cycles that Dr. Andrew Rostenburg discusses below and Rostenburg is obviously very well informed. However, he does not know about this MSM / DMSO transport pair that Gregg discovered before his death.

Watch_the **https://www.youtube.com/watch?v=BuiZa50iuKs_2015** Methylation (Cycles): Optimize Your Genes, Optimize Your Life. Presented by Dr. Andrew Rostenberg. His points:

a) Low fat and how the diet compromises your brain and body.
b) GMO harms our DNA (and China will not import our grains today due to GMO grain. The DNA damage from GMO's is passed on.)
c) You can't move carbon atoms around without the methylation cycles that turn genes on and off.
d) Why Autism is cured beyond what Patrick McGean, has reported in the video above, claims is simply sulfur. So the key is the methylation of MSM.

What Andrew fails to explain here, and importantly for these purposes, is that carbon cycles are literally "energy" cycles. So anything that increases the effectiveness of these cycles, in this case, the energy production of the mitochondria, has a profound affect on overall health of all animals (also plants as it turns out in our recent findings).

Keep in mind also that, as discussed herein elsewhere, **MSM** is 66% methyl groups and only 33% organic sulfur. Strangely, no one today, anywhere, recognizes the profound effects of MSM, but Andrew points out their value even though he knows nothing of HL MSM.

27. The Biochemical Oxygen Transport Pair (OTP) - DMSO and MSM

I include here a group of reports by the Late Dr. David Gregg, PhD and chemical engineer. His work was mainly with DMSO. DMSO (an oxidant), as he tells us, is the parent of MSM (oxidized form). Hereinafter, the two together are referred to as the OTP (Oxygen Transport Pair) (His term) with reference to OP (Oxygen Pair).

The point is that this family of organic sulfur works wonders to keep people well in any amount and he gives us more scientifically based and rational reasons as to why they are so effective beyond "Methylation" and the more basic effects of organic sulfur itself.

More importantly, I ask the reader to please grasp the fact when you ratchet the amount up, there can only be an improvement if you have not met the deficiency... Simple to understand, but apparently illusive to such brilliant investigators as Dr. Gregg's. As Sosa (above) alludes and Stephanie Seneff (also a chemical engineer) in her video posted earlier, most all of us are deficient . While Dr. Gregg did not likely grasp this key fact, he did understand the significance of these two organic chemicals and, apparently, studied them his whole life with great interest.

I start with the following point of view on MSM and parent DMSO, edited from Dr. Gregg's, "Biochemical Oxygen Transport Pair," with explanations as appropriate:

Dr. David Gregg's Theory on this OTP is that they represent an exceptionally effective OTS (Oxygen Transport System). This theory further explains some of the incredible healing properties that are reported beyond the more conventional Sulfur and Methyl components. Also, he says MSM and DMSO work essentially the same in the body and taking either one should have essentially the same effects. Then, he tells us how the two shift back and forth yielding and taking up oxygen in an incredible organic dance that has

to this day evaded most of mainstream science. Finally, in Ch.27 below, he tell us his own first-hand experience and reports as awesome testimonial on how DMSO given orally to a stroke victim completely gave her back her life. .
http://krysalishealthnotes.blogspot.com/

DMSO, as reported earlier, is a commercial solvent derived from trees as a byproduct of paper production with many health benefits derived from its use (some of which are further outlined by Bernard Caine at the end of the Group Discussions herein):

Interestingly, Dr. Gregg proposes that DMSO may provide a similar vital function in the biochemistry of plants (below). If so, it may be no accident that this OTP provides such broad ranging health benefits to us. The remarkable qualities of DMSO have been reported in many books, but one worthy of mention is, "DMSO, Nature's Healer" by Dr. Morton Walker **http://www.drmortonwalker.com/** another is by Dr. Julian Whitaker in his newsletter "Health and Healing
DMSO and MSM are both sold in health food stores throughout in the US and in most countries around the world and can be purchased online. They are used by millions of people for their health benefits. Dr. Gregg recommends DMSO for its profound beneficial effect in the treatment of Crohn's Disease and urogenital disorders. This product is commonly used for treating animals and veterinarians routinely use them for joint and muscle injuries in horses and other domestic animals. DMSO and MSM are available online as both the common liquid and as a cream with no prescription needed.
http://www.drwhitaker.com/dmso-treatment-relieves-joint-and-muscle-pain/

Read more at this link: How to Use DMSO for Healing | eHow.com
http://www.ehow.com/how_5626889_use-dmso-healing.html#ixzz2J00x3R1L
Unfortunately, the FDA has only approved DMSO for humans in the treatment of Interstitial Cystitis and Crohn's Disease. Since so many people use it for many things, this has made it a controversial subject. Possibly one reason for this is that few have explained its workings biochemically. Doing this should help raise it from the controversial "magical" health benefit classification (*although it is*) and give it a more sound technical basis.

Crohn's Disease

Dr. Gregg in his article linked below on Crohn's Disease describes how DMSO stimulates a specialized mechanism that enhances the transport of iron from the intestine to the body's cells. This is not the primary health benefit mechanism of this pair, but it further reflects their versatility. **http://www.blogger.com/profile/02628832651013377887** Much of what is explained herein is from Dr. Gregg's work.

Primary Mechanism

In his theory, both act as a "recycling system" that enhances **liver function** by improving the action of the master antioxidant Glutathione, thus reducing liver stress as noted previously. Glutathione is the path to Antiaging that Duncan Crowe so often cites. Both act as biochemistry mechanisms that enhance metabolism as explained also in the link below:
http://www.ehow.com/facts_5042811_benefits-dmso.html

Oxidation States of the OTP
Anti-Oxidants

Before we get into what Gregg says about oxygen transport (OT) and oxygen, let's discuss briefly what an anti-oxidant and an oxidant are (discussed elsewhere also): An anti-oxidant is an oxygen free radical scavenger and an oxidant burns up (oxidizes) whatever it come in contact with. As Gregg notes, this "transport pair" does both and quite efficiently... **"And importantly, they shift back and forth repeatedly"** as edited directly from Dr. Gregg's work:

Three States

In the human body and also possibly in plants, DMSO takes on three different oxidation states that can be considered in equilibrium, with the distribution between them determined by conditions that exist within the cells. All three have the properties as a group of being soluble in both oil and water-based liquids. Three states are:

1. DMSO has two Methyl atoms (M's) and one oxygen atom (O) attached to a sulfur atom (S) soluble mainly in water-based liquids, but somewhat in oils.
2. Dimethyl Sulfide is DMS with the oxygen atom removed thus with no oxygen's attached. Dimethyl sulfide is hydrophobic (soluble in oil-based liquids, but somewhat in water).
3. Dimethyl Sulfone, known as Methyl Sulfonyl Methane (MSM), is DMSO (Dimethyl Sulfoxide) with an additional oxygen atom attached (the oxidized form). Thus, this is a molecule with two oxygen atoms and it is equally soluble in both liquids.

MSM=DMSO Keeping things simple, the three are considered as two herein. The two can be considered as one organic chemical, but in oxidized and un-oxidized states. ***Actually, this is commonly the assumption in discussions in our online Groups and for all intents and purposes they should be considered as one organic chemical in all of their internal affects as he further relates below:***

28. DMSO & MSM- the OTP

The incredible oxygen transport pair.

This discussion points out just one of important contributions DMSO and MSM make in their work. It does not take into account their sulfur contribution discussed in many places here, nor that of methylation discussed in ***Ch.10. (Methyl Component of Telomere Function):***

Dr. Gregg's further findings: DMSO and MSM, form each other in the body and are **essentially indistinguishable in their biochemical effects** and that they reach equilibrium dependent on the local body chemistry and **independent of which one you use.**

The Metabolic Enhancement and Antioxidant Mechanism of the pair working together

Recognizing the above and considering that different places in the body affect the OP. If the combination is exposed to a high OP. DMSO is oxidized to MSM resulting in a higher distribution of MSM. As the combination moves to a zone of lower OP, the MSM releases its oxygen, delivering it to the metabolic processes. This results in a new distribution that is higher in DMSO again. This cycle repeats itself, serving as an exceptionally effective OTS.

This system operates on a macro scale, enhancing the transport of oxygen from the lungs to the blood, from the blood to the cells, and on a micro scale within the cells transporting high OP to the lower OP of the mitochondria where it is used in metabolism (oxidation). Review Ch. 11 to see how this affects mitochondria and Antiaging... The AHP... It all makes perfect sense.

Thus, from the above, we see a continuously decreasing OP starting highest at the entry point, the lungs (oxidation), lower OP in the blood, lower yet OP as it enters the cells, continuously lowered in OP the sequences of the metabolic reactions, and, finally, lowest OP of all as it is excreted as carbon dioxide and water... the products of metabolism.

High ORP (Oxygen Reduction Potential) + High Oxygen Importantly, in terms of Antiaging, this Oxygen Transport System (OTS) delivers critical oxygen to the mitochondria while lowering Oxidative Stress at the Liver (where we want a high ORP of glutathione) as mentioned earlier. The Oxygen Transport Pair (OTP) plays a less significant role after the oxygen has been delivered at the mitochondria and the complex metabolic reaction sequences start and this is what we are looking for in Anti-Aging and Health.

The OTS is particularly effective because all three states are soluble in both oil and water. Importantly, these are small molecules. They diffuse rapidly through both hydrophilic cell cytoplasm as well as interior hydrophobic cell membranes. This is handled in part by free electron transport (recall discussions on their adding electrical potential earlier), which can be thought of as the equivalent of Oxygen Potential (OP). The OTP have no barriers. Elimination (beyond the complex cellular process fully explained earlier) occurs not only by excretion, but also through the lungs and skin in the form of dimethyl sulfide... part of the reason that interval training exercises are so important to the AHP. There are no other known naturally occurring molecules that cause all of this to occur so effectively.

Garlic Breath Dr. Gregg, in his article, asks: Since we obtain DMSO from a plant source, do plants also use it this way? More importantly, he asks, is it an important, but unrecognized component obtained from plants that assist in our metabolism? Finally, he asks, is it an accident that the body develops a garlic smell when one uses DMSO? Could this explain some of the health benefits of garlic? Note that Garlic breath never occurs when using MSM alone. Empirically, this is nice to know, but "why" appears to be a missing question in his very clever observations and theory.

Gregg's explanation for garlic breath and the "garlic" body odor associated with DMSO: Any dimethyl sulfide formed eventually migrates to the interior of the hemoglobin molecule... the blood. However, dimethyl sulfide has a high vapor pressure so it is evaporated at the blood vessels on contact with air, in the lungs and skin. So my question is: if they bounce back and forth between molecules so readily, does MSM do this also? This may never be answered completely. But given the incredible effectiveness of MSM, we can assume that it does. Still, the garlic breath never occurs. Why not?

There is one other likely path: Transported to the lungs with its high Oxygen Potential (OP), it is oxidized back into solution, and not released in the breath. However, this conversion is not complete and can vary with individuals. My comment is, maybe this is the entire path of MSM.

Using DMSO as an Analytic Tool (Dr. Gregg's last expressed idea)
Particularly ill people have greater garlic breath when on DMSO and this diminishes greatly as they get well. Thus, DMSO could be used as a diagnostic tool to test the OP of the blood in the lungs. This could be measured, quantified, and used as a diagnostic testing tool for measuring wellness (I would not hold my breath on this one, LOL).

A Calculation on the Oxygen Transport System (OTS)

DMSO has been measured to occur naturally in our bodies (blood) at a concentration of approximately 0.2 parts per million. So we naturally make this chemistry using sulfur to a small degree.

A calculation: As oxygen is transported from our lungs to the mitochondria, it goes through the stages described above reported as oxygen pressure. In the final location, the myoglobin, the partial pressure is ca.40 mm Hg. Calculating the concentration of oxygen in water (the primary component of the cytoplasm), the partial pressure is 0.2 parts/ million, the same as the natural occurrence of DMSO. So it works!

My comments: Given the above, it is likely that DMSO has always been operating as part of the Oxygen Transport System (OTS), but has gone unrecognized. This explains why the mitochondria has so little difficulty in using it. Even my friend Craig, at 375lbs, would have only 0.04 grams in his body. Thus, a few grams added with *HL* MSM enhance a person's OT system beyond conception. This further explains why MSM works immediately to relieve back pain, most likely, which it often does. Add Gregg's observations to the cellular effects and we begin to see just how powerful this whole AHP is and why.

The Biochemical Actions of the AHP

Within an individual cell, this OTS will work to smooth-out differences in OP between organelles, lowering the potential for damage in some, while making more OT available to the mitochondria for metabolism. Enzymes, plus antioxidants allow the mitochondria to use this OT effectively through a multitude of reactions.

Antioxidants and Enzymes

For this AHP to work properly, we must include the cofactor, antioxidants and many enzymes, in our diet to compliment it. Water (and especially our enhanced "Living Water" as reported on in "Pretty Fine Sex is Spiritual," is just one source of those in this powerful Anti-Aging & Health Program. The Glutathione Conjugation Pathway is the key player in this process, so we are back to a healthy liver with the sulfur component of this OTP in this process as discussed earlier.. Enzymes, Anti-Oxidants & Liver Detoxification.

A First-Line Treatment for A Multitude of Medical Diseases/Disorders

Once it is understood that this OTP act as a profoundly effective, recycling OTS. This opens up the opportunity to use this information to treat a multitude of illnesses and problems caused by trauma that result in a deficiency of oxygen transport such as in minimizing the damage from a traumatic brain injury due to head trauma or stroke. It is well known that brain cells, more than any others in the body, cannot tolerate a lack of oxygen for an extended period. But it is proving itself effective today for any joint trauma also. This OTP will enhance oxygen transport to them until there is sufficient healing and blood flow is reestablished

29. First-hand Experience using DMSO

(Taken directly from the late **Dr. Gregg's** report)

Dr. Gregg's first-hand report using DMSO for swollen prostate:
Using jelled DMSO spread liberally over his prostate gland.

"There seemed to be considerable reduction in swelling, but more relevant to this discussion, I went to play tennis a few hours after applying it, and I felt a very large increase in energy and agility over what I was used to. I have repeated this a number of times since with the same result."

"Is this a validation of the theory of energy enhancement presented above? I am convinced it does work for me, and it is not just a placebo effect."

There is a potential for Allergic Reaction, but very few people are allergic to DMSO: Reactive sites can bond with oxygen and also bond with some protein molecules in the body, forming new molecules that invoke an immunological (allergic) reaction.

Also, numerous readers are reporting great results with MSM for prostate issues of all kinds. Some report issues subsiding within a week of taking the HL MSM. This makes a lot of men really happy as well as their wives who can sleep though the night.

From Dr. David Gregg's report relative to the use of DMSO for dementia:

10/ 7/ 03
From: Debbie
Subject: re: my Mom and memory dmso

Hi, just looking at the web sites. And... ran across your great site.
I have DMSO . . . and use for blood infection, yucky feelings . . . and yes its great and fantastic with hydrogen peroxide. Ok. The question is:

My mother is here and sent to me via Dad due to the fact he can't handle her memory loss and confusion anymore. I'm the health nut . . . smile, of the family and taking her in for 3 weeks. It's a struggle . . . she had a sudden blood clot 911 scene about 3 yrs... clot in arm... so large the arm exploded! Skin ripped open like paper tearing... and a clot in the gut as large as a melon... and the heart went haywire... ended in a pace maker...and in the brain a black spot was found

Ok, I did an experiment...Gave my mom some liquid DMSO in juice...half tsp. and within an hour...a new woman!...She could have conversation, laugh, be alert. not exhausted, stopped yawning nonstop...the look of senile confusion left. Yes, some cognitive tastes still rough...the DMSO wore off....and then back to the fatigued, yawning and disorientated person...not remember a hoot...short term memory...goes!

So, I gave some more DMSO...maybe a whole tsp....and within an hour she came alive...again...and could use the bathroom, dress her self, old conversation have energy....Oxygen must be getting to her brain,
She is only a beautiful...69...this blood clot ordeal happened...3 years ago.
If you could see her...she is so pretty...and her love for life and thankfulness just comes alive after DMSO.

Please help me....am I right for thinking this...oxygen got thru to her brain! Will this

continue to help her and even heal up the damage?
The dr. heard this story and wanted me to immediately put her on a hyperbaric treatment program. Then chelation...mouth is full of silver and gold teeth...
I hope you can help me asp. I only have my mom for one more week...I will for sure send her home with DMSO...or have Dad order from you? And is this legal in Mich. and I'm in Oklahoma where its legal...I love it personally.

Sincerely,
Debbie

10/ 7/ 03

Re: my Mom and memory DMSO

Hi Debbie,

Wow! What a wonderful testimony. It definitely makes sense in that the DMSO transports oxygen in the blood where her hemoglobin apparently isn't. She must be anemic. Read my web page on anemia: www.krysalis.net/anemia.htm

You need to correct the anemia and then the DMSO won't be needed so often. As described (the way I do it for myself), get a glass eyedropper bottle (2oz). Add (about) 10 B12, 10 folic acid and 4 multiple vitamin tablets *(not recommended here, but OK).* Fill with DMSO & shake periodically. Takes about 3 days for pills to fall apart. Let sludge settle and apply eyedropper load of clear liquid to insensitive skin.

I use the outside of my forearm. It will absorb through the skin and within about one hour she will feel wonderful. You may want to use it every day for a few days but then once a week or once a month. You will soon notice that the up feelings will start to last much longer than when DMSO alone is used. This works because B12 and folic acid are required for the body to synthesize hemoglobin and her intestine is not absorbing them properly *(Deb never read this book obviously.*

Please give me permission to add your email, with or without your name as you wish (if you want it protected) to my DMSO web page. It is too perfect an illustration to not share with others.

David
From: Debbie
10/8/03
To: David Gregg
Subject: re: Mom, dementia is OK.

What a wonderful letter. Thank you for writing. I have my Mom in bed next to me at the computer...in our office. She had a wonderful day ...on DMSO. Yes, I really think she is B 12 deficient...One day I gave her a bunch of B12 and the DMSO together...and it was wonderful.

Yes, it does wear off and I was so concerned about the wear off. My Mom and I are so happy now after reading your e mail tonight. It was so such a blessings. Our whole family is pulling for her here at my house. We love her so much. Yes, you may use my e mail to you for the website but please correct the spelling and such due to my hurriedness...at my husband's office.

I am amazed...I prayed and we stood on the fact that God wants her well...and we have made the decision to fight!...Suddenly the DMSO...idea came to me and the rest is history...that brown bottle in the cupboard was a miracle to my mother.

I will go to you anemia website now and read to my Mother. She is so happy that you wrote back and it gives such hope. This may be the answer...and not chelation?
Or...??? We will give the DMSO and B vit theory a real GO! I will stay in touch with you and hopefully...in a few days ..I will sit Mom down and let her write to you herself.
More later and may God richly bless you!!!!!!!!
Debbie
Debbie's Update-2

From: Debbie
10/15/03
To: David Gregg
Subject: re: Mom with dementia

Thank you.
I have so much to tell. Mom has been doing fantastic! She flew home and Dad was amazed at how well she was doing! I have a ton to tell. I will call you , Ok?...This is all so amazing.

Do you have other testimonials?

I have added this mineral...Germanium. Look up DMSO/germanium regarding dementia/alz.
I mean it...my Mom's life has turned around. Dad was even checking out alz. homes for Mom while she was staying here with me. I'm in Ok and they are in Mich. More later.
I will just have to talk with you. God is so good. Bless you and have a wonderful weekend...Yahoo!!!!!!

Sincerely,
Debbie (End Snip)

DMSO as a Delivery System
Finally, my old friend Craig has a lot of past experience with DMSO, but is just finding out about MSM. On reading this, he has decided to try dissolving organic sourced minerals and vitamins in DMSO and taking them orally. DMSO is very tricky in that it is such an effective carrier that it can carry toxins into your system also, by accident.

Skin Care and the OTP as a Carrier: Obviously, the options abound with DMSO especially, since it is such a powerful carrier through the skin as long as the person using it is aware that they must be careful in not allowing toxic products to enter with the chosen nutrients. However, MSM creams are commonly sold commercially and why not combine the OTP into one cream? _So I formulated an Orange Creme. It is, most likely," The most effective skin cream on earth, hands-down. "MSM/ DMSO is the "Delivery System. Please do not inquire as I do not sell it commercially.

Now add: aloe vera gel, vitamins A, D, E, P5P (B-6), methlcobalamin (B-12) and B-5, Almond Oil, Jojoba Oil, Rosemary Oil, and Extracts of Chamomile, Ginseng, Grapefruit and ginkgo biloba, Cocoa Butter, Acetyl L-Carnitine, Coconut Oil, Egg Shell Membrane, Hyaluronic Acid, Red Palm Oil/Alpha Carotene, Resveratrol, Niacinamide and Astaxanthin (hence the orange) These nutrients go straight down with the aid of MSM/ DMSO and immediately go to work in healing at levels never before seen. It works as envisioned. So now, you take a daily vitamin and nourish your skin, all in one step. But you must make your own.

30. Downside of the Transport Pair

After about twenty years of using MSM and getting feedback on HL MSM and DMSO, It appears that I have finally isolated an actual problem with this therapy, but please bear with me here. There are still some issues with this discussion, but my story pans out based on reader's reports and my own personal experience as I outline herein:

To begin, many readers have complained of leg cramps when using MSM in high levels and that is my experience also, but not generally, and there are interventions. Moreover, this is a problem that goes well beyond the "bad taste" issue that is readily avoided with swishing.

Dropping back to my worst leg cramp ever, in my thigh muscle, occurred about ten years ago when I was supplementing high levels of the very efficient calcium supplement, Ezorb. Keep in mind that, at that time, I was also taking about a half cup per day of MSM. When this occurred, I attributed the entire problem to the high level of calcium absorption alone. That incident was bad enough to keep me from even walking normally or playing tennis for a week. On researching calcium at the time, I was certain that I had isolated the problem and never supplemented calcium again. Generally, as you will see below, I was mostly correct.

The new problem today: I was applying DMSO in large amounts with regularity to a one inch patch or so, a skin issue on my right leg calf. With this, I started getting leg cramps only on that calf alone. First, minor, but the last one I got was a major one, nearly off the scale, and nearly equal to the above thigh cramp from the Ezorb.

Reports of leg cramps with HL MSM are fairly common, but relationships are difficult, if not impossible, to establish cause/ effect. It is generally agreed that magnesium causes muscles to relax, while calcium tightens (contracts) them electrically. Therefore, magnesium is considered a great method of avoiding heart attacks and it is generally agreed that it is a primary food for doing that. So the answer here is cacao powder, which is about 50% magnesium. When you are getting enough cacao, there are no leg cramps.

Methylation Pathways

If you follow the methylation pathways and here is a good example of the diagram: https://images.search.yahoo.com/search/images;_ylt=A0LEV1JW9yZZGQsA745XNyoA;_ylu=X3oDMTEycTdlYjQzBGNvbG8DYmYxBHBvcwMxBHZ0aWQDQjM5NTBfMQRzZWMDc2M-?p=melylation+pathwys&fr=ytff1-yff28#id=10&iurl=https%3A%2F%2Fs-media-cache-ak0.pinimg.com%2F736x%2Fea%2F3d%2F6e%2Fea3d6e9942e3caea0fd36ac7a25d8bb2.jpg&action=click. However, following them is not an easy task as they are the most complex of microbiologic relationships known. Still, follow me here, note the methyl cycle NOS>Arginine>Ornithine> UREA. Keep in mind that a urea imbalance in the muscle is one culprit attributed to cramps. Obviously, if MSM/ DMSO enhances the cycles as I have proposed based on the many healing results seen in the reports, this could be at least a contributing factor as I established here with my DMSO experience.

However, beyond that, my further experience is that with adequate magnesium, with oral HL MSM, cramps do not normally occur. Thus, as part of my protocol now, I recommended the supplement Cacao to help offset any magnesium imbalance. This is a tasty, rich source of organic magnesium and it can be used in many creative ways.

While the methyl diagram (address above) is plenty complex, it does not include a transformation that I am fairly certain occurs constantly as follows: The body, in its magic, can convert calcium directly to magnesium (and back to some degree). This is a radical theory but it tests out. You are not about to read it in any scientific publication or even on the web today and this is only based on my own observations. If this is so, and you get adequate magnesium as with Cacao. You should never supplement calcium or concern yourself with getting it into your diet.

I have recommended against calcium supplementation for many years and this new theory goes a long way toward explaining why you just never should do it. Obviously, calcium supplementation is very popular with the mainstream and it is cheap. Today, it is put in many forms of food from orange juice on, and in all daily vitamin pills.

So my bottom line here is that magnesium, especially any natural form, such as cacao is a huge addition to any diet. However, keep in mind this premise that magnesium is a two way street and a healthy body will readily balance the two. When you have muscle cramps then, that cramp is a warning that the overall balance is lost at least in this local muscle. Further, a high intake of DMSO as with my calf cramp could set the imbalanced conditions off and actually be the cause.

Dropping back some: A more mainstream observation is that we balance our calcium levels by stealing it from our bones, hence, osteoporosis occurs when we have an inadequate supply of calcium. With the above in mind, an inadequate supply of magnesium is a direct cause of bone wasting that all have managed to miss.

Thus you hear from mainstream sources everywhere of how virtually all women and most older men suffer from bone loss and indeed we do see that older people grow backwards. However, just to give you hope: Note that I immediately went back to my lifelong height (about 1 1/2" gain) when I started the HLMSM protocol and I am still holding that height today at age 75. This experience helped initiate the birth of the Mg>Ca (and back) premise that I report here today. Add the considerable joint pain and relief from HL MSM and the above cramps and you have my inspiration here for the above theory. How many doctors even suggest bone regrowth (and it could all be cartilage, I just don't know yet)?

So what precautions can you take to offset using DMSO topically in large amounts? One obvious solution would be to apply liposomal magnesium cream or liquid to the site in conjunction with the DMSO. One source: https://www.amazon.com/Liposomal-bis-glycinate-Formulated-Seeking-health/dp/B00BJO3YJ0/ref=sr_1_1_a_it?ie=UTF8&qid=1495727694&sr=8-1&keywords=liposomal+magnesium

I have not personally tried this thus far, but the logic is there. The key idea here is to balance the minerals in an area at a rate that exceeds the body's ability to work and thus keep the muscle's relaxed and comfortable.

DMSO continues to be an amazing topical treatment, especially when used with others in that it drives them in just as Liposomals do. I prefer MSM orally, but DMSO is also a great mouthwash to heal gums and it has great potential in many ways as a simple healer.

31. Pharmaceutical Drugs

On a more general level, the current 2.6 trillion dollar "fad" that started, maybe a hundred years ago, has led us to the synthesis of drugs that can be patented and sold. Drugs and Allopathic Medicine (conventional science based in the Newtonian Science ca.1670's) are strictly aimed at dividing the magical parts of life into their components parts and their being patented and controlled. More specifically, when it comes to disease, it is about separating a disease into its various effects and treating the effects, rather than the root cause. In rare cases, this can actually work, but generally, the results are ineffective.

With pharmaceutical drugs, it is usually about using a product of nature of a derived chemical. From nature, for instance, consider White Willow Bark. The Romans found its value and Bayer separated out aspirin. So what is considered to be the active ingredient was separated out, then patented. The results are thus less effective, but proprietary today.

The very idea of separating the various biological factors that reward life into separate entities is counter intuitive. In fact, when this is done, the phytonutrients and cofactors are lost and the result is a crass image of the original herb that results in the many reported side effects common to drugs generally. Today, Bayer is a synthetic version, not plant derived. But then, if the drug industry did not have the attitude that they are the anointed ones, we would not have Bayer as a player in their industry and we would still be doing what the Romans did. The question is, what have we actually gained in our quality of life from this? White Willow Bark was being used as a drug even then.

Nature's Intelligence, the science that Newton gave us has changed very little in 350 years. These chemicals that nature gave us in its mystery and wisdom will never be understood even if one appears to stand out as "cause." The magical part that scientists are looking for. In some ways, they must always destroy, but in other ways they can enhance an organism to some degree even though parts are lost. Generally, however, the original intelligence was far superior to the synthesized form and the best that we can do is just bottle it and sell it as a substitute for good nutrition and a sound understanding of nature..

Modern allotropic medicine's effort, with the exception of trauma care and a few others, is like painting a ceramic tile counter a new color. The paint may stick and it may even look OK, but underneath the original colored glaze remains. Left alone very long, the paint will flake off... with what results? Herein, we are interested in fundamental changes that

last. When these occur, everyone can see us Thrive as we age more gracefully.
http://www.youtube.com/watch?v=8JF7TcPsmvI

Side effects: By design, as pharmaceutical drugs must always be foreign to an organism in some way, meaning negative side effects. In the case of packaged foods, given that they are foreign to nature, by design increase our chances of our being allergic to compared to organic holistic food. This too is integral to the AHP formulae. (5th edition)
http://www.anh-usa.org/durbin-anti-supplement-backfired/

Allergies: All of the cures mentioned here must have "side effects." However, in our case, the side effects are positive. You may have or may develop allergies even with certain natural cures, but keep in mind how allergies start (discussed earlier). This tells you that they can come and go, but with proper nutrition and liver conditioning, they need not be permanent.

Treat yourself well: As established science tells us, the body replicates itself about every sixteen to twenty years (opinions vary). But everyone generally agrees that this occurs. By treating your body well, the next version may quite likely be stronger and have fewer or no allergies. We are learning how to treat it well here. However, most people don't know where to start and are misguided by commercial claims as profit margins drive them to new drugs and their next version will likely be worse than this one.

32. The Demonization of Saturated Fats

Transfats are now understood too be the basis of our increasing heart disease according to even the most mainstream of scientists today. When you hear the story behind it, you begin to understand just how far behind the FDA and Drug companies are from the real science. So below is the background story:

The Lipid Hypothesis

This was first developed to explain the pathogenesis of atherosclerosis by the German pathologist Rudolf Virchow in 1856. In 1913, Nikolai Anitschkow's study showed that feeding rabbits cholesterol could induce symptoms similar to atherosclerosis, suggesting a role for cholesterol in this disease. Did anyone think this through? It does not take a high degree of wisdom to understand that rabbits are not meant to consume animal fats. Even still, the lipid hypothesis caught on to some degree as a result and remain in the mainstream even today.

The Framingham Study

The 1948 Framingham Study involving 5,209 people aged 30 to 62 is well known as a source of support for the lipid hypothesis and today remains the gold standard of proof of the theory that eating saturated fat will raise blood serum cholesterol and lead to cardiovascular disease or CVD.

https://en.wikipedia.org/wiki/Framingham_Heart_Study
Unfortunately, much of the data was misinterpreted and remains so, according to Dr. Joseph Mercola, but it does not take an MD to see the broadly misinterpreted mistakes. In 1953, Dr. Ancel Keys published a paper comparing saturated fat intake and heart disease mortality. The Cotton industry, about that time, was looking for a way to sell Cotton Seed Oil (CSO). CSO, as the name implies, was a byproduct in the production of cotton. I am told that it's an ugly gray, gravy-like material. My relatives all raised water melons and cantaloupes, so I never had the pleasure of consuming it, but it sounds lovely.

The dairy industry in those days was a powerful lobby and dairy farming was then extremely lucrative. However, the CSO people wanted a piece of their action. So they started pawning off yellow colored, fluffed-up CSO calling it margarine as a healthy alternative to butter and, using the lipid hypothesis, began their marketing.

This did not sit well with the dairy farmers of the time, so they stopped the food coloring in its tracks, declaring how vile the stuff was (and rightly so). Well, it's hard to stop a creative marketer and the cotton seed people obviously had a few ideas, plus money. So they put the yellow dye in a separate container alongside the ugly fluffed-up gray stuff. Then they discovered the Keys report and enlisted Keys to help them sell margarine as a healthy alternative.

It was apparently easy to sell people against the fear of heart attacks, no matter how baseless the findings. The word quickly got out everywhere that the saturated fat in dairy products caused hardening of the arteries and ultimately led to heart attacks. One issue of Life Magazine even featured their hero, Dr. Ancel Keys (who had degrees in economics, political science, oceanography, and PhD's in biology, and physiology (but was not a medical scientist or M.D), on its cover. What a great start!

The writing was on the wall. Most housewives mixed the yellow dye and the fluffed up cotton seed till about 1958 when everyone, including the "experts," were thoroughly convinced that the dairy industry was killing them. As the word got out, the market share for margarine grew with leaps and bounds and dairy industry nearly completely lost its power... Which is pretty much where we stand today as our FDA allows Key's conclusions to be marketed as God's word and most mainstream doctors still follow this.

If you "Can't Believe its Not Butter," you are right, but it unhealthy and not as good tasting as butter (unless you can't tolerate butter, then you should try the coconut or cacao butter, both great substitutes. Nevertheless, by 1978, most researchers had bought into Ancel's theory by a wide margin. By 2002, the lipid hypothesis had become "scientific law" where it stands for the most part today.

By the way, as an aside, Dr. Ancel Keys is also known for his "K-rations" (Keys rations) consisting of hard biscuits, dry sausage, hard candy, and chocolate, which should give you some idea of his real knowledge of nutrition. The military ate it up, literally. Maybe this had something to do with why all of the sergeants that I knew in the Air Force died so quickly after retirement while still in their 40's. **http://en.wikipedia.org/wiki/Ancel_Keys**

To quote George V. Mann, Sc.D. M.D. Professor of Biochemistry at Vanderbilt University School of Medicine.
http://www.people.com/people/archive/article/0,,20072775,00.html *In an editorial printed in the American Heart Journal in 1978, he wrote: For 25 years the treatment dogma for coronary heart disease) has been a low cholesterol, low fat, polyunsaturated diet. This treatment grew out of a reasonable hypothesis raised in 1950 by Gofman and*

others, but soon a lot of aggressive industrialists, self-interested foundations, and selfish scientists turned this hypothesis into nutritional dogma which was widely impressed upon physicians and the general public (so this was likely well known forty years ago).

A nadir was reached when zealous doctors and salesmen arranged such "prudent" meals for national meetings of cardiologists, rather like Tupper-ware teas. They grew up in the interface between science and the government funding agencies a club of devoted supporters of the dogma which controlled the funding of research, a group known by the cynics among us as the "heart mafia."

Critics or disbelievers of the diet/heart dogma were seen as pariahs and they went unfunded, while such extravagancies as the Diet/Heart trial, the MRFIT trial and a dozen or more lavish Lipid Research Centers divided up the booty. For a generation, research on heart disease has been far more political than scientific. All this resulted from the abuse of the scientific method. A valid hypothesis was raised, tested, and found untenable. But for selfish reasons, it has not been abandoned.

To those who are paying attention, the Dr. Ancel Keys theory has turned out to be just as ugly as the cotton seed oil story, but the misguided ousting of saturated fat from the suggested mainstream diet has continued ever since. Fortunately, the truth is finally starting to come out, as medical scientists have finally begun to question Keys' findings. The truth is that margarine, vegetable shortening, and partially hydrogenated vegetable oil/ trans fats... are the true villains. But they become far worse when heated to higher temperatures. Remember this when you hear, "Fries with that burger, Sir?" And they are causing a multitude of health problems other than just heart disease as kids now are getting Type II diabetes as reported later here.

One of the driving forces behind the FDA allowing the keeping of the lipid hypothesis has been the blockbuster statin drug industry. Without that theory, statins make no real sense and, literally, a half-trillion dollar (so far) blockbuster drug goes bust. When and if that occurs, the entire drug industry could fall flat on its face and right behind them, possibly, the AMA... who has supported the theories for many years. So a great deal rides on this sham. Should this occur, it could open up the kind of creative medicine that Dr. Max Gerson practiced in the 1950's and encourage talented physicians like my friend, Dr. Kevin Freeman to go back into the profession. But government would have to change greatly before that could occur and the mechanism for that is difficult to conceive.

Dr. Ernest N. Cutis, M.D. Quote *"The Cholesterol Delusion"*

- Highlight Loc. 1196-99 | **http://www.amazon.com/Cholesterol-Delusion-Ernest-Curtis-M-D/dp/1608447480**

The pharmaceutical industry obviously has a huge stake in the survival of the Cholesterol Theory. It is estimated that statin drugs alone steal $20–$30 billion per year from the population. Add to that the cost of the equipment, chemical reagents, and other paraphernalia for testing not only cholesterol levels, but also the liver and muscle enzyme tests required every few weeks, and you get a total expenditure on this one class of drugs that exceeds the gross national product of many countries around the world (in all, statins are, likely, a $Trillion drug and now we know that they have no beneficial effect and are in fact, very detrimental overall just as lowering saturated fats in general is).

Dr. Uffe Ravinskov, MD

Further refs: Cholesterol is not a deadly poison, but a substance vital to the cells of all mammals Uffe Ravinskov, MD, PhD **http://www.ravnskov.nu/cholesterol/**

Also read: "Does High Cholesterol REALLY Cause Heart Disease?" **http://articles.mercola.com/sites/articles/archive/2009/12/05/does-high-cholesterol-really-cause-heart-disease.aspx** by Uffe Ravinskov, MD, PhD and The Cholesterol Myths **http://www.ravnskov.nu/cholesterol/**

Coconut Oil:

All saturated fats including the most healthy of all known oils, Coconut Oil, naturally fell into the category of a bad fat for heart health as a result of Keys' paper. Coconut Oil is mostly of medium chain lauric acid triglycerides (LAT's). If you have any doubts about the healthy qualities of these LATs, you should realize that LAT's are also very prevalent in human breast milk. It can also curb hunger as it helps to provide a very lasting energy source for anyone including babies. But then, Keys saw candy as a major food group when he invented K-rations, so we should not be surprised that he made them villains.

We consume as much coconut oil as we can in our house and, it, alongside red palm oil and Cacao butter http://n.stuccu.com/s/Cacao+Butter (also known to contain LATs) is the most healthy of cooking oils. Coconut oil is the ideal fat to use when you are seeking to replace calories from grains and sugar (Carbs), when on the AHP. As a general rule, no cooking oil, no matter what it is, should ever be heated over 140 degrees F internal temperature (so no deep fried foods). 140 deg F is the point where virtually all enzymes have been killed. We need all of the enzymes that we can get, plus we can actually store enzymes to some degree as mentioned earlier. The higher our enzyme stores, the better, always. also read **https://en.wikipedia.org/wiki/Cocoa_butter**

MCT Oil

Medium-chain triglycerides oils (MCT oils) are used in special medical formulas for people who need energy from fat but have trouble digesting it from regular dietary sources. Great for cooking because they impart no flavor in foods as Coconut and Cacao can and they are readily absorbed and used by the brain, etc. They can be a solid solution to dietary fats.

Dr. Leonard Coldwell

As a spokesman for the AHP, it would be difficult to find a closer agreement, generally, than with this MD. Listen to this interview and he will verify that cholesterol levels of 250+ are the most healing and how green foods are oxygenating... and basically how cancer works. But, importantly, at this juncture, he verifies that Ancel Keys was wrong, by his recommendations and Keys' must be vilified or people will keep dying.
https://www.informationvine.com/index?qsrc=999&qo=semQuery&ad=semD&o=36162&l=sem&askid=4433cd72-4a4c-4f33-a7d2-988d295a32b1-0-iv_gsb&q=dr%20leonard%20coldwell%20cancer&dqi=&am=broad&an=google_s

33. Genetics... Generally Insignificant

The mainstream loves the highest form of technology possible. This removes you from "their" game... they demand control... But it is your body:

Dr. Danny Hillis

Listen to Dr. Danny Hills of Applied Minds http://appliedminds Ted Talk on Proteonics and Cancer
http://www.bing.com/search?q=dr.%20danny%20hills%20of%20applied%20minds%20ted%20talk&pc=conduit&ptag=G407-AEC48904830B84430ABF&form=CONBNT&conlogo=CT3210127&ShowAppsUI=1
You discover that diseases, especially what we generally consider DNA defective disease are mostly the result in a defect in the conversation between proteins, now called Proteonics (and certainly not a part of or a defect in our gene pool) as Dr. Bruce Lipton has pointed out elsewhere.

The mainstream always wants to tell us that our health problems are insolvable defects that only can be resolved (masked) by drugs. They have convinced most of us that they are correct and thus are able to market a 2.6 trillion business smiling with confidence. To this, I suggest that we simply leave them to rot. Further, that the world would be far better off if these people and their health related support groups simply retired and played golf eight hours a day, rather than worked.

Genetics and Healing

So what part does genetics play in the healing process? Generally, none, but again, the mainstream would have you believe that when you begin to show outward signs of cancer, it has nothing to do with our diet or environment as discussed herein. They want you to believe that there is little that anyone can do for cancer other than slash, burn and poison. They are killing us just as they killed my lovely brother, Steve, who edited the 1st edition of this book in 2016 and for whom this 4th edition book is dedicated. He was pushed to genetic cancer drugs and lasted a year. I miss him, his wonderful love for life, and his great sense of humor ... one of the great people in my life of consequence.

Personal Responsibility

It is about money and control and it is part of their bent in eliminating personal responsibility from anything in your life. Give up all control and they will be gladly help you. So our society hates to point fingers and people learn to never take blame for any problems or conditions, whether physical or mental health or what have you.

Problems: Johnny is failing because he has bad teachers, Suzie caught her cold from the kids in school. We went broke because the market failed... The stories are endless. This is not to say that outside conditions don't affect us, but if we must take control of our lives, We must admit to ourselves where the responsibility lies. This is always our dream, and our game, that we are dreaming. We agree to and own them.

When did we hear: I have contracted cancer or had a heart attack because I have failed to get the proper healing nutrients in my diet that promote glyconutrient and Proteonic cellular communication. But when you finally hear something akin to that, you know that we have arrived. Of course, cancer is caused by any number of complex events, but it is clear that we can affect both its occurrence and cure. I predict that boosting Dietary Protein, Fats, and Sulfur levels together with some of the other clever cofactors will prove to be the next area that Hills and his group will find to be a major component in the curing of cancer as they organize their giant think tank on Cancer and related diseases.

Most all alternative healers have moved in the above direction and are moving further that way as time goes on. Many still discuss the genetic component, but that component is considered as not over 10% of any given cause. As result, the alternative community is growing with leaps and bounds as the mainstream falters. Eventually, they will pile on, but we must first make them realize that they are serving us and not the other way around. **Shirking** personal responsibility makes you a spiritual infant... and doing it sets you up for many, many more problems down the road. You either face those problems as challenges and solve them as brother Steve tried, or dodge them. This is gonna take awhile. Furthermore, as I have said a several times here in many different ways, life on earth is not meant to be easy, but it is our responsibility to attempt to make it better as this book is attempting to do.

Problems are meant to be resolved by you and no one else. No, that does not mean that you can't seek help. Actually, asking for help can be the start. It may indicate that you are not a block head... or worse. We are not alone on any spiritual level and help is always available if you ask. The earth is a spiritual place even though this fact is often nearly impossible to recognize. But, back to the above point, there is a great deal of difference between asking for help and shirking responsibility.

Masters: Highly spiritual people (masters, and yes there are masters on earth) are often impossible or difficult to recognize. They may sense your shortcomings and even deliberately anger you, for valid reasons. While they are here to help give you a spiritual boost... this is your world to conquer and rise above. They will seldom push you for any reason.

Change: It makes sense that a spiritual boost may make you very unhappy in the short term because it means change ... and we all inherently hate change. So it will not be fun, but your job as a spiritual being is to maintain a loving attitude and recognize who they are, just as you love the people who humor you or make you happy for their own personal gains.... the salesmen of the world. Salesmen know how to make you happy, but happiness is not necessarily what you need to take the next step spiritually.

Healers As I mention later on, we are all blessed with at least one healer in our lives. The trick is always about recognizing them when they are here. They can often slip away before you understand what is occurring. If that occurs and you later realize who they were, I can guarantee you that you will be one sorry Soul. What a shame to be given a spiritual gift and not recognize it, but we all do that. The good news is that you will always be given another chance. Unfortunately, that opportunity may be years away. So the key here is to be vigilant and aware. Life is indeed a spiritual journey.

34 Beyond Genetics (Epigenetics)

Intestinal gut floras capture our attention. Bigger than genes and maybe less complex, they influence everything to a much greater degree than assumed previously:

Microbiaone **http: //en.wikipedia.org/wiki/Microbiome** is a relatively new word to all of us. It describes the new found ecological community (flora) of outside organisms that cooperatively and presently inhabit our bodies and it is generally applied to the digestive tract. However, we also have them in every orifice, plus the skin.

In the Air Force, the guys would joke that they were not taking a shower because they did not want to wash off "the protective germs..." a joke then, but now we see the truth in it. These protective germs exist in spades. No, I am not recommending never taking showers, but the new antiseptic hand soaps are a really bad idea unless you are a surgeon.

Over the last ten years, the alternative health community has placed more and more importance on biotics. Literally, you carry pounds of these tiny creatures. That is, 100 Trillion in a single human. They inhabit our digestive system and cooperate with our internal defense systems to keep us well and allow us to Thrive. To think that only thirty years ago most people did not even know that they existed. We thought that we were independent organisms and all bacteria needed to be killed as a problem. This is a mind set that many still have that started with Louis Pasteur and all commercial products still advertise **http://www.biography.com/people/louis-pasteur-9434402L**
As with many discoveries, good things can be taken to extremes as with this example.

An Old Memory Returns: Our selective memory of past events is always interesting, a story comes to mind: I can still clearly recall a valuable inner lesson that I received one night in the dream state in 1956 as an 8th grader. The morning after that dream occurred, I clearly recall relating it to my Mom over breakfast before school. I described a vast community of "good germs" that inhabit our body and keep us well. The dream was about sixty years early compared to the science of that day. However, as I read about the Microbiaone, that dream plays back. As a result of what I had learned, I was less concerned with eating carrots out of the ground at Grandma's... and today, likely as a result, I am never affected by eating disorders, asthma, or allergies with the single exception that fiberglass makes me itch. My sensitivity to fiberglass began when building my 26' sailboat in 1970, but it took a lot of fiberglass grinding (with no protection) to make that occur. I still can't inspect a house that is being insulated with fiberglass without itching.

Back to the story, unlike what most parents would have commonly done in those days, Mom listened intently. You could tell by her loving composure that she understood the importance of my lesson. She always encouraged me to tell her what I had learned in my dreams... what I had been given access to... as I explained that we had all of these "good germs" (probably the only description that I had of them) in our systems that protect us and keep us well. I can still recall how amazed I was with my story, myself, that day. It was as if I were an outside reporter relating a news story that had been handed to me. That is often how spiritual lessons are recalled, because we really are often out of our mental bodies (minds) when they occur.

Today, as a result of the internet, this information is out there for all of us to study and take advantage of. There are keys to protecting this flora that include prebiotics (like inulin, below), or the natural material that occurs in soil and plants that help influence the feeding of this flora and probiotics, or food that they can be taken that will help maintain them.

35. Inulin & Moringa (5th edition)

Inulin is a vegetable component that aids in promoting the Microbiaone. Moringa is a tree that can do the same, but is more general:

Inulin Our discussions about bowel health would be remiss without at least a bone thrown to Inulin. Inulin is a natural sweetener and nutrient that appears in nearly all vegetables to some degree, but it is derived as a supplement from chicory. It can be problematic for a few individuals, through, causing overgrowth of intestinal methanogenic bacteria... which produce gas. Inulin increases organic calcium and magnesium absorption, so it has star qualities if you are not one of the unfortunate ones who cannot use it. Like MSM, add it gradually to your diet. **https://en.wikipedia.org/wiki/Inulin**

Moringa Tree: One natural nutrient that anyone can grow, at least as a seasonal tree, that I discovered a some years ago is the Moringa Tree. This tropical tree from India can be grown in most locations. It grows 2" a day with almost no water and in marginal soils. All parts can be ingested, but the leaves and roots should be particularly interesting to the LCD people. I now buy the leaves and tea online, but the plan is to have a grove of them some day. Also, moringa leaf powder can be bought online for about $15/ lb. Including S&H if you shop carefully. Some of the truly incredible properties of Moringa:

Liver Health, first and foremost: Moringa, being a powerful liver nutrient, counteracts liver damage from alcohol and repairs mitochondrial damage caused by excess alcohol intake, plus it can help repair pharmaceutical drug damage.

Anti-oxidant Power: The highest ORAC (oxygen reduction coefficient) of any known food ever tested by anyone. As an anti-oxidant, it has no peers.

Published Moringa Oleifera benefits claimed (some sound a bit crazy, but this stuff is outrageous):

- **Vitamins and Minerals**: With over 90 known nutrients including 46 known antioxidants (so far), 7x Vit C of oranges. 4x Vit A of carrots, 4x calcium of milk, 4x potassium of bananas, 3x Vit E of spinach and the list goes on.
- **Protein Source:** Contains ALL of the essential amino acids, so is **a complete** plant protein source... not the case in any other known plant. Plant proteins are always deficient in one or more amino acids... meaning they are incomplete proteins. Plus it is rich in poly phenols, carotenoids, & flavinoids.
- **Bone loss Prevention:** It must be an efficient way to guard against bone loss for post menopausal women among its many applications, given that it is so high in plant-based calcium. Calcium is incredibly difficult to get from a safe organic source with no intolerance or allergy issues noted and this is balanced with magnesium, so no cramps or the other negatives associated with calcium.
- **Alkalizer:** It naturally alkalizes the body and balances the pH, so it is a wonderful natural cancer fighter.
- **Phytonutrient Power House:** Contains more phytonutrients than ever measured in any fruit or vegetable. It is a natural diuretic, anti spasmodic, anti bacterial, anti fungal, and is a Cox2 inhibitor so it inhibits prostaglandins... a major concern with heart problems.
- **Antimicrobial:** Guards against systemic infections.
- **Anti-inflammatory:** The most potent Anti-inflammatory & natural pain reliever ever tested.

- **Eliminates heavy metals** by stimulating the formation of metal binding proteins for detoxification... and it works on all three levels. It works like MSM in this way.
- **Mainstream Testing:** Has been extensively tested by the NIH as a healing source for disease prevention... and, if you know about the NIH (National Institute of Health), you know that "they virtually never look at foods as a healing source" but this one is suddenly getting everyone's attention even though it was actually mentioned in the Bible.

My first-hand experience with Moringa and **How it works:**

- It significantly boosts my energy levels when I add one to three teaspoonsful of the dried leaf to my coffee/whey mix in the morning (described later). Improves my tennis game by increasing my endurance and reflexes. This is a natural increase and there is no peak. **No Colds:** I have never contracted a cold since adding this food to my diet some seven years ago.
-

Beyond all of the above, it is dirt cheap to buy compared to most supplements that do far less. A pound lasts me two months or more. However, you could probably just grow this tree alone with less work and a higher level of health than you could by growing a complete garden... and subsist on it. But a warning: it can cause detox... loose stools.

In tropical climates, people all over the world grow it in their yards with no special attention. In Florida, one tree will produce leaves and berries (which look like bean pods) for years on end. It currently is being grown as a cash crop in tropical climates world wide on plantations where other crops will not grow. Also, it grows quickly in any warm climate, however, even in moderate climates, it will act as a very productive annual. The taste, is mildly spicy, but it tastes great and it adds zest to cooked meals. You can buy the seeds online.

The dried leaf powder is more potent than the fresh leaves! This sounds crazy, but it's true... and the list goes on, but again, do not overdo this stuff! Its is powerful. Go slowly!

Michael Pollan as he relates in his book (referenced below): Virtually every farmer in the U.S. had a family garden in the '40's. It was a necessity. Of course, any garden grown then was produced using manure from the pasture mixed in with humus soil that included the stalks and leftover parts of the previous crops. This almost never happens today.

This may not have been a Caveman Diet, but it is a vital component of the LCD... You must get plenty of fresh multicolored vegetables or you simply are not really doing the program. It's a given that without these basic organic nutrients, any organism can't Thrive.

36. Antibiotics

The Microbiaone killer drugs and related health problem:

The very most important aspect of the above Microbiaone story is really not that we all have this mass of cooperative organisms in our guts, but that they are easily killed off by antibiotics. Most doctors don't get this yet today. They hand out antibiotics like candy and we are causing huge problems with certain bacteria becoming resistant. My close friend, Dr. Rich Price's wife, Joanne died of MRSA, a severe case of this.

Antibiotics must always be general, so they always kill the good guys. This is a growing concern and also could be a killer in ways that few doctors currently consider. Eating candy has a similar effect, but no one ever ate candy in an effort to become well (with the possible exception of the commercial chocolate scam). However, any healthcare professional who writes a prescription for an antibiotic http://sustainableagriculture.net/ and does not accompany it with a stern warning to take every last pill and does not include a prebiotic and probiotic in his list should have his license revoked, IMHO.

CAFOs I am neither an animal activist nor an environmental hawk. Nevertheless, despite the above growing problems with prescription dispersed antibiotics, our greatest source of commonly received antibiotics is not even concerned with prescriptions, but with Large-scale concentrated animal feeding operations (CAFOs) according to the National Sustainable Agriculture Coalition, "Many CAFO operators give antibiotics to animals to make them grow faster and prevent diseases that are caused by the extreme crowding and other stresses on the animals. An estimated 70 percent of antibiotics and related drugs produced in this country are used in animal agriculture for these non- therapeutic purposes." http://sustainableagriculture.net/our-work/conservation-environment/clean-water-act/

The Omnivore's Dilemma Michael Pollan relates in, " The Omnivore's Dilemma" **http://michaelpollan.com/books/the-omnivores-dilemma/** that CAFOs could not function without antibiotics. According to Woodgate's view eating meats grown in factory farmed conditions creates probably 80% of all free antibiotics that are released into our environment, particularly into our rivers and streams. If this does not give you pause, nothing should. Our US meats are long term killers as a result.

http://woodgatesview.com/2011/05/17/antibiotic-use-by-cafos-increases-our-health-care-costs-3/

We don't yet know yet how much, if any, of the antibiotics are destroyed by cooking and processing the meat, but you can bet that we are about to learn as these bell ringers come out to tell us how modern farming is killing us in small bytes and bits.

Could anyone have predicted where this would lead when the AMA destroyed Royal Rife and implemented this modality? Electronic healing could have taken the place of all antibiotics had Rife been given the backing. We have proven that this is a non-invasive and cheaper therapy. Even though the FDA frowns on it, it is, nevertheless, commonly employed in the alternative health community, but never in animals.
The chances are not so good if you live in a big city, but if all possible, you really want to befriend a local farmer who can sell you their meat and dairy... I did that in Boston. When you do it, you can actually check out their operation and then you know, first-hand (remember, first-hand knowledge always rules). Dairy is less a culprit than cattle.

Earl Warren My people are Earl Warren and his sons (until 2015 his late wife, Mary Anne). I have learned from Earl that a small farmer can actually make more money selling product that avoids the factory farming pitfalls... and you will mutually benefit, if you happen to live in the Rehoboth, DE area. His operation is called "Rehoboth Dairy Farm" and he actually sells whole milk in returnable glass bottles. It took me years to find him, but you must find someone like him if you expect to live beyond the lifetime that our average population is headed in total wellness and Thrive. Visit them at: http://www.aboutmybeaches.com/16186/rehoboth-dairy-rustic-acres-farm-century-farm-grass-fed-cows-no-artificial-hormones-no-antibiotics-the-way-milk-should-taste-amish-eggs-butter-breads-rehoboth-beach-delaware-beaches/

The die out of Microbiaone as a result of antibiotic overuse has been linked to a range of nasty conditions, including obesity, arthritis, and a myriad of other conditions, but most likely we have just scratched the surface on this one. I can't imagine why I would have been given this information so early if this was not the real key to wellness. Currently, two newer areas of research are pushing this field further. The possible gut/ bug link is to a pair of very different conditions: autism and irritable bowel disease, but again, this is the tip of the iceberg.

The AHP is key to getting there, but the one single pitfall that is most disturbing and that should alarm you is the antibiotics, especially in meat cattle, but also in chickens and pigs. (Sadly, I must report that in July of 2015, Mary Ann Died of Cancer Treatment).

37. Digestive Enzymes

For those of us who have learned that supplementation is really the only path to wellness, given what we know about farming practices, etc., I offer the following:

When you take tablets and pills, these are devoid of the enzymes that naturally occur in the fruits, vegetables, and meats that would normally be your only staples as a "Cave Man." Therefore, you must supplement digestive enzymes with whatever it is that you take.

Bromelain One of the most effective and advantageous enzymes to supplement is Bromelain. The effects of using it have been known since 1875 as an enzyme, phytomedical compound and anti-inflammatory. Bromelain is derived from the high stem structure of pineapples. The part that most people throw out (not me). For an excellent study on Bromelain, go to: http://www.ncbi.nlm.nih.gov/pmc/articles/PMC3529416/

Bromelain Facts: On reading this study, you discover that Bromelain is not just a great digestive aid, it could also be classed as a miracle supplement in its own right. Interestingly, you could take as much as 12 grams of it with no adverse effects and doing this can be effective to:

- Cure cancer/ tumor regression in vitro and in vivo (Read the report. The effects are impressive)
- Reduce angina pectoris and the effects of heart disease in general.
- Reverse and inhibit of platelet aggregation (thick blood)
- Minimize the risk of arterial thrombosis and embolism
- Aid against sinusitis
- Reduce surgical traumas and reduce the number of days of trauma after surgery
- Treat disorders of the blood vessels and heart, coronary heart disease
- Ward off stokes and reduce the effects of a stroke.
- Reduce peripheral artery disease (PAD)
- Reduce the pain and effects of both Rheumatoid and Osteo arthritis and is an

effective alternative to NAISDs
- Reduce the effects of autoimmune diseases

There are other digestive enzymes such as papaya that are also effective, but none have the added benefits that Bromelain yields. It is an inexpensive supplement that should be taken with any capsule and before meals, when possible. Digestion, in itself is an energy consuming event that reduces our capacity to maximize the nutrients in our foods. When used effectively, Bromelain, along with digestive flora can maximize the process.

38. Water Solutions and Health(4th edition)

Interesting ways to make water magical:

Water, as we all know is essential to life and especially wellness. As I report below, Dr. Dan Nelson's "Wayback" water was a huge step in the search for an ideal solution, Also, at one time, I saw Megahydrate as the pentacle of water products. Megahydrate helped me greatly to control migraines for years in my own wellness search, but now I do not use it as I report on the "Living Water" device below. Migraines no longer occur after 74 years. Dr. Price is selling these right now in limited quantities. He has 300 on hand for $600 each. Email him at RichSPrice@comcast.net Try calling 302 393 4523, but he is a busy man and we have yet to get a marketing program in place. The device is a single pass and it includes a pass-thru water filter. So it is a two-step process, but quick. He can set you up with a whole house system if you are inclined.

Migraines, as I spoke of earlier, were a significant problem for me to overcome and I had still had not completely won that battle until very recently. But once I found Wayback, they were not even close to the problem that they were at, say, age14 when I would go blind, somewhat deaf, and my brain would basically freeze. Then, my body temp would often drop to 91 Degrees F... A range where the heart will sometimes stop and death follows. Hence, migraines can be deadly.

On Observation, what follows is completely counter intuitive, but true: During my military years, I found that extreme physical exercise/ activity made migraines subside. It's not something that a person who has them can normally fathom, but if you get your heart rate up high enough, the attack gets "lost." It comes back some eventually, but at least this provides relief. I used my combination bike for this over the years, so much that it was completely shot... I had to rebuild it as a result and reminder. But these attacks pointed out the remarkable characteristics of one of my personal, first-hand discoveries, MegaHydrate and then the healing qualities of these more complex water solutions like Dan's and finally, our latest one below, "Living Water." .

Ion Induced Water Changes (5th edition):
http://pubs.acs.org/doi/abs/10.1021/acs.langmuir.5b04489

Hofmeister salts (salts of essential minerals) alter water structure. Sea salts naturally contain these ions and thus can potentially change the chemistry of water, thus produce RacA, a naturally aggregating protein that could be used in DNA repair.

Obviously, this experiment is preliminary, and inconclusive, but should open your eyes as to what Dan Nelson has done with Wayback and finally, what Dr. Rich Price is doing with Living water when you have adequate minerals in your diet. These are all vibratory in nature using infrared spectrum vibrations and thus can change the very chemistry of biology.

Dan and Rich's work is proprietary, but this is a published 2016 experiment with plenty of data to back it up. This will alter how things grow as we have seen with our "Living water." Combine this with Epigenetics and you have the natural means for DNA repair that transcends anything current science even considers possible.

Megahydrate and Crystal Energy
http://www.wetterwater.net/why_take_crystalenergy.html and "Megahydrate"(then called MicroHydrin) were by Patrick Flanagan and represent only a small part of the many hundreds of his inventions. His ideas are still, literally, outer space things. However, he has managed to have escaped mainstream recognition even though he has apparently also done well commercially. You can read about him and his many wild inventions at the above site.

More on ORP Some further background on understanding what we are discussing: **ORP** commonly stands for oxidation reduction potential. A high (positive) ORP reading indicates a strong oxidizing potential. For clarity, we will use **OP** to mean a high Oxygen Potential (positive) and ORP to mean oxygen removing potential or a measure of its free radical scavenging. Both OP and ORP are important to keeping us well as Dr. Flanagan has elegantly pointed out in his talks. Oxygen will literally burn up outside invaders and free us of disease (with the help of white blood cells and fever's (high body temperature) aid). In this process, the free radicals created are savaged by antioxidants or we become poisoned by waste oxygen (debris, if you will).

So, Oxygen keeps us alive and Anti-Oxidants scavenge the by-products... put another way, oxygen burns our food and helps kill invaders while antioxidants and the liver (producing glutathione) keeps us young. Dr. Flanagan's water additive, Crystal Energy and his tablet powder, MegaHydrate both concentrate on the ORP side of this equation... While *HL* MSM and Whey do both, added antioxidants can be a welcome insurance policy in this overall AHP. We want to keep the waste side as clean and free of oxygen waste products as possible.

As Dr. Flanagan basically explains in his talks, oxygen, the key to life itself, is by its name, the oxidizer... oxidation can be thought of as rust... and aging can be thought of as rusting. Thus, oxygen is both important to wellness and life, while it is the aging factor when it is not cleaned up. So, without the powerful help of the antioxidant Glutathione and some assistance from such nutrients as Resveratrol and MegaHydrate in this complex process, we just rust away into old age... and that is what most people have come to expect to do. We have a better way.

So here is a first-hand story of the astounding effects of Crystal Energy that Patrick sells at http://www.megahydrate.com/ I watched the video tape that Dr. Hobbs gave me after her personal episode and ordered both products. Crystal Energy is used by placing two drops in a glass of water as Dr. Hobbs did. It reduces the surface tension of water and makes it far more absorbable (by the way, a four ounce container can last for several years).

However, I, myself, did not notice any immediate benefits from it even though it seems to make water taste better. Nevertheless, as Patrick will tell you lowering the surface tension of water allows the cells to hydrate. It improves the "communication" between cells. I used a couple of drops in just about everything that I drink for that reason. By the way, fresh fruits possess these same benefits in their juices and when you juice them and keep the juices in a container, they disappear. Therefore, we can assume that the effects are both electrical and chemical.

Now moving onto his powder, which is very different: As a result of hearing Dr. Hobb's account, when I tried MicroHydrin. I quickly recognized that it would stop a migraine headache in its tracks within a few minutes. Now that was a find! The migraine would come back eventually, if I did not jump on my exercise bike, but I was blown away by its effect. I had used and recommended it to everyone till Living Water was invented... from migraine sufferers to stroke or spinal injury victims...it can have magical effects where body electronics have been possibly short-circuited.

Megahydrate MicroHydrin, in a proprietary way, alters electrically signals between our brain and our organs when combined with water... and it does this almost immediately, on ingestion. Given what it does for migraines and knowing the theories behind it, this product is nothing short of otherworldly in its ability to positively change our basic bodies. If you read the works of others here and research them in the links that I have given, this relationship becomes Crystal Clear (pun intended).

If you begin to catch a cold, that is, you have not kept up your D3 levels, take three or four tablets (along with 50k iu of D3) immediately and it will give you a natural quick boost, increasing your anti-oxidant levels exponentially. It works by helping you maintain your electrical signals between cells, and improve communication between organs. Do that a few times and there will be no cold.

MegaHydrate (MicroHydrin) is nothing short of a chemical solution to a Beck Device with added, truly magical, benefits. From here, we have the most incredible water ever:

Living Water(5th edition) Enter the newest solution to my friend Dr. Richard Price's arsenal noted above. Eventually, this will not simply be water and it is related to the above referenced abstract. This is a water pass device now in limited production. I have tested it as a single pass device for people for about a year and the results are phenomenal in terms of health, As I report elsewhere herein and one might suspect, it is working on plants for irrigation as well as on animals and people. It is truly phenomenal and health changing. More on the most recent experiments with plants in, "Pretty Fine Sex is Spiritual."

39. The Beck Device and Royal Rife's Machine

Dr. Robert Beck As members of the Beck-blood-electronics Forum, I would be remiss not to discuss Dr. Robert Beck's work. While Flanagan seems to have created electrical healing chemically, Bob Beck undoubtedly did it well using electrical currents and electromagnetic healing in various ways. Years before, as well known to those in our group, Royal Rife apparently killed bacteria and viruses using harmonics/ electronics and observed their deaths under a microscope that could pick up a living virus.

Bob Beck, no less brilliant than Rife, showed us that he alter cell structure with just a 9 volt current or using the photographic flash that he invented. Some in our groups have reported cures that could have killed us or otherwise left us with unmanageable conditions. His work has changed the quality of lives and extended them in undeniable ways even as science has ignored what he did. You Tube has videos of both of these amazing men available. So going further with observations here is a waste of time. But if you are truly interested in healing, you should watch all of Beck and Rife's videos.
http://www.youtube.com/watch?v=AysfKyl8O9k
http://www.youtube.com/watch?v=yC7M_rhg2yU

40. The Anti-aging and Spiritual Component of Healing. (4th edition)

System Failure equals Disease While a biological organism is seen as a series of combined car-like parts by most medical scientists, we are missing a great deal when we attempt to "treat" a disease as anything but a temporary or permanent overall system failure. Modern medicine takes the problem apart and treats each symptom separately and, nearly always, each part of an organ, separately. For example, surgeons do a bypass surgery when our liver has become overloaded and too much plaque builds up in our arteries. The plaque itself is likely a life extension measure imposed by our body to sustain it due to our bad diet, but it does not occur when our diet is optimal.

When we treat the flu, we treat the symptoms, acting against our natural defense systems; We take aspirins and Tylenol to lower fever. So why do we get fevers? They kill viruses by raising body temperature in addition, the heat also stimulates your body's immune system to a heightened response. We take decongestants to stop our noses from running. These and others occur as a part of our defense system. While uncomfortable, they make you well. Your body generally knows what is best... remember this when buying over-the-counter drugs. Finally, we see our doctor and he gives us an antibacterial to guard against a "possible" secondary infection depriving us of natural gut bacteria.

There are always exceptions though to the above, as a wise doctor reminded me years ago, when my son was running a very high fever from an ear infection. Children, especially, can run high fevers to the point that their bodies just can no longer handle them and they must be lowered or they can be permanently brain damaged. Of course, ear infections are the result of many problems that can be avoided and antibiotics are always a temporary solution that beg recurrence (not to mention a defensive medicine practice, legally).

While an infection may show up and appear in one part of an organism, it is always systemic in nature, even if it may not be obvious. An extreme case is the parasitic heart worms in a dog. I have spoken to dog owners who have had dogs that have actually survived heart worms. Heart worms, you would think, would always be fatal, but not the case. So, even parasites and fungal infections are evidence of a degree of systemic breakdown. From the other viewpoint, resistance to all infections are a measure of our ability to Thrive.

Systemic Cure By introducing drugs, we have actually robbed nature of its ability to find systemic cures. While we may not be willing to accept this level of harsh reality, its true. Organisms increase their ability to thrive by overcoming infections through their own internal or externally adapted defenses. For this reason, roaches and rats are doing far better as a population than people and our protected pets are. Remember, they never visit doctors, vets, or hospitals and they are literally Thriving.

While an advocate of Anti-Aging today, my bent is in living a productive, pain-free life at as high a level of awareness as possible. In the process, my intention is to add to the experiences of others with books like this and help them achieve just what it is that they are looking for as they progress spiritually. There is no wrong or right here. You just move up as you gain experience. Strangely, don't feel stupid if you think that you have not gotten anywhere spiritually, because if you feel that way, you probably are further along than most. At least, you are aware that spirituality can occur... a start.

One other thought regarding how spirituality affects our physical bodies. As some scientists have discovered, as harmonic vibrations increase, our physical bodies become more well... and some have found that the electromagnetic parts of this are quite measurable. However, if we become too far advanced spiritually, we just don't need a physical body anymore... and guess what? We abandon it! So death, itself, can be an indication of spiritual achievement.

Spiritual Adventure This physical world is very efficient, but the higher worlds are much more so. Even though the physical world appears as anything but spiritual, it is the very basis for the entire system. Nothing is wasted here even though it appears to us as though there is waste everywhere. We use the terms: wasting time, environmental disaster, illusions, etc., but these are all actually teaching devices. Correctly viewed, life on earth is a spiritual adventure.

Four Physical Laws So, the point is that when you are ready to move along, you go. You abandon your body immediately, without a whimper or a complaint, despite the fact that the physical body wants to and even could, live forever. This is a basic spiritual fact, but below are the four other basic tenants of this Earth Planet... physical laws, if you will:

- It is a harsh place and will remain so. It is designed to produce a high level of difficulty, but we are here to learn and this is our training ground. In fact, it is "perfect" in that sense.

- All people must eventually achieve higher awareness. However, in the process, they can only hear what they want. So here, no matter what or how often the truth is revealed... until their awareness level is raised through experience mainly through hardship and love can be used. Until then, people turn a deaf ear to it.

- When a person's level of awareness rises, they then can grasp more truth. It's as if, when they are ready, their hearing improves (even if they are going older).

- This entire world is constantly moving up in awareness as time goes on. We all will eventually become spiritual masters. A corollary is that as you learn more love, it spreads and we all move up spiritually together... Nice thought, huh?

Everything in this world is tied together... And when a group or an authority that we abide by attempts to tear it apart as did Hitler or the USSR in the past or the AMA and the FDA today, they are violating spiritual laws, but they are also serving a spiritual purpose even if it is unclear. Still, eventually, they must in turn, make payment and re-balance the situation. In the end, it all balances out. Right and wrong are mental illusions.

41. The Subconscious and Beyond (4th edition)

Bruce Lipton - 'The Power Of Consciousness" Watch: **Https://www.youtube.com/channel/UCtSQfqA-Sv3DMScpk32vbIg**

Dr. Bruce Lipton has been a great contributor over the years to pure genetic science, but now he leaps into the new frontier of Epigentics and the Sub-Conscious Realm.
He was cloning stem cells forty years ago and now has moved into what really counts. The environment controls cells, not genes. Genes do not do what we are told. They do not

control life. Cancer is not controlled by genes, you control your genes.

His book, “The Biology of Belief.” How you control your unconscious mind.
http://www.thriftbooks.com/w/the-biology-of-belief-unleashing-the-power-of-consciousness-matter-and-miracles-by-bruce-h-lipton/251173/?mkwid=sheYWuUrg|dc&pcrid=79077774672&pkw=_cat:thriftbooks.com&pmt=b&plc=&gclid=CO347qC-99EC_FZqCswod810ARw#isbn=0975991477
Thank you Bruce! Here goes and I really needed a well established scientist like Bruce to make my points here. This is in no way a religion and in some ways, it is just the opposite. That is, these higher levels are divorced from established belief systems. No one invented them and each individual must establish their own methods that are often unique to them.

I love this stuff, but this is a related topic in a sense. After all what does the sub-conscious have to do with MSM and Supplements? Actually, plenty, once you understand the overall picture, but it is not so obvious. Whenever, you look at it, as Bruce has, and I always have, its really just one topic. So if this sounds redundant, it's because it is, in a way. Therefore, as Bruce points out in his video, the key to getting to the sub-conscious level is through conscious awareness. Know your stuff. The sub-conscious will grasp it and give you back more than you bargained for.

Sub-Conscious Level Here goes: One thing that we seldom discuss, here or anywhere, is the Sub-Conscious level of healing. However, science is proving that raising our spiritual level has huge implications in our overall physical health and wellness. Do you really need someone to tell you this? Probably not, but there are thousands out there doing it... because it is true and because they think that they have, “The Way.” In fact, I have a way that works for me, but you must find yours. We are individual Souls with physical bodies that we must maintain to stay here and raise our levels of consciousness to the higher levels. The earth is our training ground and is very good at performing this function.

Morgan Freeman in “Through the Wormhole”, talked about this for almost an hour in a recent show. Their show would lead you to believe that the subconscious level is the highest obtainable. However, I am personally aware of much higher levels. These are the levels that most scientists would dismiss as noise, even if they were privy to them. But some of us actually go there in total awareness. These are the unlimited worlds of God.

Others do this also and it is not within the confines of any religion. As a result of personal experiences on these higher planes (or levels) going back to as early as age 3 (as mentioned elsewhere herein), I, long ago, dismissed death and time as an illusion that we all experience as an agreement, not fact. As a spiritual being, you are totally free and unconfined.

The Hindus were correct when they told us that the physical world is an illusion in time, just as Einstein and others later reinforced. In fact, it is all happening at once, but our minds separate it into a linear timeline. So, as previously mentioned, we don't die, we merely transform and move on to the level of awareness that we're ready to accept. Therefore, anti-aging is, really, not the point, but awareness is everything.

Even the Hindus never have quit fully understood this final part of the puzzle after ten thousand years of study. So, I simply don't care about time. The illusion of time persists until we learn how to rise above it and then it simply stops. It all stops. You can learn how to stop it using this exercise below.

Love The ultimate "place" is quite simply a dimension that I call Love. Yup, God really is love. Crazy stuff, huh? It's a nowhere place where only three things occur: love, light, and sound (all vibrations). Don't worry, you'll love it, because you are love also! It is simply an agreement in vibrations where you become totally aware of the love that permeates. It is you and you are It. You know that you have arrived when the other, more coarse, dimensions are gone. You might ask, how does someone perceive of a place or dimension when there is no time, space, energy, or even matter? I can't answer that and my mind can't even fathom it, but I can clearly recall it, even if it still it makes no sense. You are not encouraged to accept any of this on faith, in fact, no major world religion does. However, you will eventually get it and as I have said previously, you will love it, Finally, there will be no argument when you realize that it's true.

5th Dimension Call it the 5th dimension if you wish (not the music group, but that is a good start because the sound is very beautiful on this plane or level of awareness), but then, there is even more! Best of all, no one must die just to get there. Many have it backwards and this misconception causes them a lot of unnecessary grief, but we all get it. We do and we absolutely must. The system is perfect. Religions are built around the descriptions of those who have been there like Jesus and Mohammad, but since the people who want to replicate the same thing cannot possibly get it. Our paths are individualistic in every case. Since they are not there yet, they mess up the stories badly. For this reason, all written world religions are flawed.

There are not many things that I am totally certain of, but this last statement nails it: The more you express love, and the more your religion talks about love, the closer you are to truth.
This is not primarily a discussion on spirituality. But healing and wellness must be recognized as a part of this thing that we call, "Ourselves" or "Soul." We are connected to all things just as birth and death are connected to life. If you watch the Freeman program, you can't help but see that the writers could not escape this fundamental truth even if they have never apparently known about anything beyond the subconscious level of awareness.
Some links to higher teachings:
http://www.thewayoftruth.org/default.htm
http://www.eckankar.org/

By the way, my book, "Beyond Epigenetics" is about what Bruce teaches in his videos as viewed from the higher realms that I discuss above and in the following chapter.
https://www.bookdepository.com/Beyond-Epigenetics-MR-James-R-Clark/9781515060451

42. Body-Mind-Soul Connection & Contemplation

We are Soul; spiritual beings inhabiting a human body, our responsibility is in caring for it just as you take care of your car:

Soul-Body Connection Many people today in the health care community are especially aware of the Mind-Body Connection and much has been written about it. I take this notion a step further and suggest that you can actually learn to separate these, not join them, as so often taught by others.

When we learn how to separate them, we learn how to separate ourselves from situations that would overwhelm others and make them ill. So while Soul is a unity and the body is a unified organism... and we must have a body and mind (at least) to live here, they exist as a cooperation, not as a single entity. This may sound as though it is a contradiction to the AHP, but this contradiction is a paradox (in that the contradiction reveals a greater truth). This idea can be further explored in "Through the Wormhole," as narrated by Morgan Freeman

In his series, we can see how alternate areas of the mind "light up" and turn off according to where we place our attention. Despite what the writers thought, this literally occurs as we move through the various higher worlds and cross each barrier or plane. Also, each plane has its associated sound and color (vibration).

Active Mind My interest here is in Freeman's discussion of how turning off the "active mind," turned on (or lights up the "inactive-mind or subconscious-mind.)" When we go into a Contemplative State... that is, we put the active mind to sleep, we turn on the higher state as the scientists observed. From a healing and wellness point of view, doing this yields huge physical rewards beyond any physical thing that we can do to our body... Even AHP.

Space Time While the scientists on the show could measure and actually see this working, it is unlikely that anyone involved in these experiments had an idea that their work was actually just the beginning of an endless journey. My further discussion here is directed toward the mechanics of how to not only raise your conscious level, but, also, while doing that, turn off outside influences and allow rejuvenation of your physical body as in the Contemplative Exercise below. When you have mastered this, the last step in this process is to become totally aware of yourself as Soul, completely separate from both the sub-conscious and physical bodies. In this state, your body is completely left on its own to rejuvenate.

Switching Topics: One thing, on a more mundane level, that really bothered me about the above mentioned "Through the Wormhole" show was that they apparently used CT scans to pull off these brain images. CT scans produce 700 x the radiation of a chest x-ray (per 2/13 Consumer Reports) and they were doing multiple scans on this person in the experiments. From what I could see, they were done repeatedly and in a short period of time, but I have no way of knowing for sure.

The subject of these tests likely got as much radiation poison as a Hiroshima occupant by the time the procedures were completed. Most hospitals still use CTs according to a recent Consumer Reports column.

Your first choice in imaging should be an MRI, IMHO, and even those high levels of magnetic radiation are, quite likely, very dangerous. The final choice is to stay well and avoid hospitals completely, if you want to thrive. Hospitals are just not places conducive to optimum health.

You can make a grand, sudden jump into the highest of the spiritual worlds. There are no set ground rules except for a loving attitude must be there for this to work. You must develop this to Thrive.

43. A Contemplative Exercise
A method of keeping it all in tune:

As Soul, we all have infinite access to the state of awareness as I propose above. Given this, the first step is this exercise to set aside about 15 minutes a day to Training Soul in a Creative Exercise.

In our busy lives, this is asking a lot, but it can be the most rewarding of all things that you can do, ever, to help yourself in total rejuvenation. By some accounts, this could amount to as much as 50% of wellness. I suggest that it's more like 95%. But who is counting? This is free and it will change your life if you do it.

So first, find the time... Then find a quiet place. We must find a quiet place so that we can hear the inner sounds that raise within us as Soul. So we want very few noises that might interrupt this process and certainly no TV or people talking. Some can learn to actually use external sounds & thus aid in this, but at first, it is safer to start in a quiet, sacred place.

Find a comfortable, but supportive chair and relax yourself mentally. Think of those lights that the scientists observed going off in your brain and the higher ones that they saw, turning on. Find a soothing word to repeat to help calm your mind and relax. See yourself as a happy Soul, a light, in a higher realm away from this crazy, mad world. This may be a real place or a fantastic place that you build-on somewhere else in the universe.

Stay here for as long as you wish, but 15 minutes is a good exercise. Slowly come back and greet the day with a new, more loving attitude as you venture into this world of wellness and love and move up the ladder to Thrive. There are no set ground rules except for a loving attitude must be there for this to work. You must develop this to Thrive.

44. Healers
Spiritual gifts that we are all given:

We have all been in the company of a healer. In many cases, our greatest example was someone dear to us who had no formal training... maybe it was our mother or grandmother...

or even a son or daughter. In fact, a person can be very high on the spiritual ladder and still be very young in years. Also, you can be far more wise in terms of knowledge level than you are working from than they are. That is, a six year old or a sixty year old can hold this ability. What matters is where you stand relatively in terms of love. Love is the foundation of a healer. There may be other factors involved, but this is the key to spirituality.

Healers are easy for us to recognize if we are paying attention, because they are additive and never subtractive. These great beings are the ones that can help us Thrive and they are invaluable. If we have been allowed to flourish on this earth for any amount of time, we have been provided with a Healer at some point. It's a given and it's our birthright.

Allowing Healers are brought here to give you space and help you Thrive. When you find one, never attempt to hold onto their coattails... But graciously follow them when they suggest your doing so. Give them their space and they will assist you in healing yourself. Notice the wording, because this is always how it works. You are always the primary mover, the one in control.

All things in nature work as a part of the whole and healers are one of its most wonderful examples. They exist in at least two realms at once and, thus, defy time. They are quantum entities of the highest frequencies. In watching them work, the idea is to learn how to emulate their love, not to take advantage of their outer skills. If you remember to do this, your rewards will be great.

In my case, my own Mother was my healer and I will cherish her always. She never pried into my life and never asked much from me. I only hope that you can recognize and recall your spiritual Healer at some point if you do not now.

A spiritual master is always a Healer. They go hand in hand, but they are not the same thing. When a person can look as you, recognize that you have a problem, and help you get through it, you have found one. Cherish them!

For more on this topic, read "Surrender to the Oneness"
https://www.amazon.com/Surrender-Oneness-James-Robert-Clark/dp/1519499345/ref=sr_1_7?s=books&ie=UTF8&qid=1486322672&sr=1-7&refinements=p_27%3AMr.+James+Robert+Clark

45. Great Health Recipes

Having fun with the AHP:

So what is a great health recipe? A recipe is a combination of foods. We have all tasted combinations of foods that we like. If you have gotten this far, you probably agree that all recipes should include an eye on our health, but here are my criteria:

- Do you love its taste? Generally, as a complex organism, your body recognizes the things that make and keep you well. So it first must taste good to you and not even necessarily to your friends or family. If not, it misses the cut immediately.
- Fortunately, you are not trying to please a general audience, so this part of your job is easier than it is for, say, commercial chefs. You can prepare the best tasting food on earth to suite your taste relatively easily and their best would not even be in the running.
- Is it increasing your overall wellness factor? Again, having gotten this far in the reading, you are on the road to becoming a master at knowing what all of this means. You know that no one else can fill-in the empty blanks for you. However, this could take quite awhile, even if you are very good in determining what is best for you. All of us are just not starting in the same place. However, if this all rings true to you, your chances are very good that you will be Thriving soon.
- Few commercial chefs worry too much about the wellness factor in restaurants. In fact, commercial chefs leave that concern to the health department, mainly. The profit motive drives their preparations, but with your vision, we can change all of this. My vision: "Demand it... and it will come."
- Have you ingested it with the vision that it will improve your well being? Our perceptions are undeniably influenced by preconceived notions. For this reason, every commercial chef knows that presentation is important to anything that they concoct. So they all have this one down well. Anyway, make your presentation interesting if you can.
- Did you arrive at the greatest and best version? Tastes change with any number of influences.
- Every chef knows that presentation is important, but where was your attention when you tasted it? This one may take awhile, but it is yours alone to work out. This is a trial and error thing, so make sure that you have gotten it correct before you abandon it completely.

On occasion, we have had members of our group ask for our personal recipes. Once I sent in mine and a member piped up that my tea recipe was unhealthy because I had added hot water, killing the enzymes. Well, I had not specified the temperature. Adding boiling hot water to tea may help it taste better and, if it is already dead, nothing is lost. It's a fact. But the keys are:

- Make sure that if you are looking for your result to supply beneficial enzymes from some source and that you not use over 140 Deg F water in the preparation once they are present.
- Always use care in pouring the correct temperatures together also. Mix up the cooler stuff, like Moringa tea in one cup and slowly add a hotter component to it.
- Enzymes always occur with fresh organic ingredients and even with some dried ones like Moringa tea. So always use care and awareness in your preparation and blending,
- If you mess up, don't throw it out. Drink it. This is not a contest.

I make many different kinds of hotter and colder drinks, but my favorite hot drink is decaf Coffee/ Whey and Moringa and cacao. Here are "my" recipes:

- **Coffee/ Whey:** Make up a mug's worth of hot coffee no caffeine please anymore. In an empty mug, add a heaping scoop of UWPP and stir in enough cream to make a paste. Add a heaping teaspoon of Moringa leaf and stir in. Add a tbl spoon of cacao, four drops of kelp/iodine & stir some more, then any other tablet based supplements. Very slowly add your coffee stirring it in till the mug is full (if it clumps up, you went too fast)(rule: pour hot coffee into cold whey mix to reduce shock). Drink it & make some more but stir slowly this time. Save the rest separately and add that back in to keep the temperature up to 140 deg F as you drink it.

Now this is some great stuff IMHO and it tastes great cold, room temperature, or at 140 deg F. I'll finish off a cup after it has been sitting on the table for hours and relish its flavor. Moreover, it is enough food to last me till lunch when I have my Moringa/ Green Tea and it will supply enough lasting energy to play two hours of hard tennis and then some. When I finish, I am neither hungry nor tired. It's a powerful energy drink to be sure... Not a fake commercial sugar, high caffeine, concoction that commercial marketers call a sports drink. Sometimes, I'll deliberately leave ½' or so in my mug and finish it off after tennis... the Moringa in the bottom is hot/ spicy and delivers a nice lift.

Lunch for me means sardines in water maybe with a few sea salted saltine crackers followed by some citrus fruit if I don't do a whey/fruit/protein shake

Moringa/Green Tea:

- Boil a cup of water and pour over a cup of Salada Pomegranate Berry tea. Allow to simmer.
- To this add some Crystal Energy described here earlier. Realize that the hotter the water, the easier it is for it to pass through your mouth membranes, but why test it? The stuff is dirt cheap and you want this absorbed as soon as possible.
- To a half cup of 140 deg F water add a Moringa tea bag. Allow to simmer.
- Mix a tbl spoon of Moringa leaves and 1/4 cup of heavy cream and stir in a large mug.
- To the mug, slowly stir in the Moringa tea followed by the Salada.

Now this is my idea of a refreshing winter drink... Moringa is tart and it adds spice to any drink as long as you don't go too far with it.

Smoothie My favorite lunch, a whey/fruit/protein shake/smoothie, consists of:

- Putting about ½ c. of cream and ½ c. pomegranate juice in a blender: Add frozen blackberry/ blueberry/raspberry mix, a tbl spn of concord grapes that have been blended together seeds in (from my vineyard),
- Half of a frozen peach, a fresh, unfrozen orange,
- Tbl spn of frozen ground green coffee beans. Be careful to over-ground them.
- Then put them thru a fine sieve or you end up with a mess.
- Save any grounds that don't clear.
- Add a small amount of frozen pineapple (for digestive enzymes) and an organic raw egg or two... Blend all of this for quite some time.
- Add a heaping scoop or two of UWPP and lightly blend. Be careful not to overdo this or you kill the protein in the whey.

Freeze what you don't drink of this smoothly in a plastic container and reuse it or eat it as a frozen ice cream snack... It's delicious!

46. Transdermal Nutrients

I spoke of Dr. Mark Sircus' Trandermal Magnesium Therapy in Ch.20. Magnesium is a

key element in wellness... but there is more:

A Direct Path By combining nutrients into a body and face cream (and I discuss this in several Group Posts later on), we can with the help of DMSO, MSM, and (QP) Quantum Programming, cause the nutrients of choice to go directly into the body with pre-determined results that were previously unheard of.

It is Efficient! The advantages to this are that they become more than twice as efficient and responsive. Since supplements must go though the digestive tract and digestion, in itself, is maybe 30% efficient. Some foods are more efficient and are absorbed even in the mouth like MSM, which is incredibly efficient using the *HL* MSM Protocol While others just pass through, and we get very little, like magnesium. So there are extreme problems with certain nutrients. However, most foods are well handled as we might expect, since we have evolved through time as successful organisms. Of course, the AHP is about maximizing that success and this is one obvious method.

No Digestion The point of Transdermal Nutrients is to bypass the digestive process completely. On the negative side, we must use extreme care in doing this, since we are also bypassing the body's defense systems. In fact, nearly everyone does this every day, women especially, when they apply lotions, creams, deodorants, lipsticks, salves, etc.

Precautions What we place on our skin is generally absorbed straight into the body and sent directly to the liver. Many times they are not well handled, thus, we compromise our health. Without question, some of these unregulated beauty products result in cancer and death, since they can introduce excess aluminum and heavy metals into the body. When this occurs, there can be no explanation given, but the results are obvious and easy enough to avoid if you read the labels carefully.

Positive Benefits So back to the positive side, because this delivery system is so effective, we can put safe nutritive products directly where we need them and quickly. Taking advantage of the incredible abilities of DMSO and MSM as outlined in Ch. 26 by Dr. Gregg as delivery systems and adding a proprietary QP, we can make up a great range of very effective skin and organ multipliers that would otherwise never would be delivered in effective enough amounts to the create profound results. This is especially true as we age, since our digestive systems lose effectiveness over time. However, our skin does not. So no matter how old you are, this system never loses effectiveness as a delivery system.

Toxicant Removal and other Benefits Finally, following the sage advice of Jon Barron, we can literally rid our bodies on a, say, four month schedule, of toxicants by placing Cilantro and Chlorella Tincture in the cream, applying it for a couple of weeks and then going back to the basic formulae. Furthermore, when we put Astaxanthin in the cream, we have a visual method of determining that it has been absorbed. Finally, Astaxanthin, itself, protects us from the harmful effects of solar radiation while allowing us to spend virtually unlimited time in the sun (read my earlier story in Ch.23 on this) and still not be sunburned. I am not recommending this, but we all get "stuck" sometimes, just as I did at the U.S. Open and we must get our natural source of D-3 when we can. With this formulation, then, we get some rather incredible both health and antiaging benefits that commercial beauty creams could never deliver and at the same time avoid the commercial toxins that they contain.

Preparation for Using Dermal Abrasion or Exfoliation is a way to get rid of older skin and bring the small capillaries closer to the surface of the skin. When you do that, you allow transdermal applications to become more effective. There are many ways of doing this, but first, even though your mother told you not to... When you itch, scratch it. In fact, scratching is a natural way of helping rid your body of old skin. When you introduce nutrients, it is even better. Of course, you can overdo anything, but keep this in mind. Furthermore, there is a skin care product called a Buf Puf that is available in many stores and online at:
http://www.amazon.com/dp/B0015GQZ2U/ref=asc_df_B0015GQZ2U2405301?smid=AO44X8DCPV0V0&tag=sdcbing77-20&linkCode=asn&creative=395117&creativeASIN=B0015GQZ2U

Buy the Buf-Puf with reusable handle on it (about $6.50), so that you can reach places in your back and legs without effort. Use this product "Green Side Up" only, so that you get maximum exfoliation in a hot shower that opens skin pores prior to applying transdermal magnesium and creams on your entire body and face. Never use products with DMSO or MSM on unwashed skin. This is especially true if you use any beauty products even if you are positive that they do not contain any toxic chemicals. While they do not... Ask yourself the question, do I want lipstick or eye shadow injected into my blood? If you have bothered to have gotten this far in this book, we know that the answer is absolutely not.

However, a final word of warning: If you make up your own cream, be cautious of what you include because: "It's the Liver Stupid."

47. The Killer Cosmetics

Since I ventured into producing a natural body cream recently with the idea of making a cream that functions as a nutrient delivery system for Anti-Aging and Health, I have done research on products to add as well as, importantly, products to avoid and why. Dr. Mercola also did a piece just recently on commercial shampoo and the dangers attached to using these preparations. Some, like Dioxins and Sulfuric Acid would set off alarms with just about anyone with a rudimentary chemical background, but others are far less recognizable. Here are some primary examples:

Sodium Lauryl Sulfate (SLS), sodium dodecyl sulfate,C12H25NaO4S http://www.flexyx.com/A/Akyposal%20SDS.html) is a surfactant, meaning that it is used to break down fats and oils. Also called Aquarex ME (see Sulfuric Acid), Sodium dodecyl sulfate, Sodium salt sulfuric acid, Aquarex methyl, Akyposal SDS, and host of other pseudonyms.

- According to the industry, SLS has a long history of "safe use" in a variety of consumer personal care products. Others dispute this vehemently, as our cancer rates increase on a yearly basis and we look for sound reasons.
- It is used in toothpastes and shampoos Toxicity @ Overdose: ORAL (LD50): Acute: 1288 mg/kg [Rat] is a foaming agent derived from coconut and/or palm kernel oil (So far, so good.).
- It is used in shampoos, creams and pastes to properly disperse their ingredients.
- So now we get into the negatives: It is probably the most dangerous ingredient used in skin and hair-care products according to most healthwatch groups.
- In the cleaning industry SLS is used in garage floor cleaners, engine de-greasers, car-washsoaps, etc. It is very corrosive and readily attacks greasy surfaces SLS is used throughout the world for clinical testing," As a primary skin irritant."

- Laboratories use it to irritate skin on test animals and humans so that they may then test healing agents to see how effective they are... Are you getting this?
- A study at the U. of Georgia Medical College, indicated that SLS penetrated into the eyes as well as brain, heart, liver, etc., and showed long-term retention in the tissues. The study also indicated that SLS penetrated young children's eyes and prevented them from developing properly and caused cataracts to develop.
- In adults: SLS in shampoos may actually **cause** hair loss by attacking the follicles.
- It is actually classified as a "drug" in bubble baths because it eats away skin protection and causes rashes.
- Harmful to both skin and hair, it cleans by corrosion. It dries the skin by stripping the protective lipids from the surface, so that it can't effectively regulate moisture.
- Another potential serious problem: SLS reacts with many types of ingredients used in skin products and forms nitrosomines (nitrates). Nitrates are considered cancer-causing carcinogenics by many, but this may be incorrect.
- Because of the alarming penetrating power of SLS, large amounts of these are absorbed through the skin into the body. We just don't know enough about them yet to know what this means at this point. If they are carcinogens, this problem is big.
- A variation of SLS is SODIUM LAURETH SULFATE (Sodium Lauryl Ether Sulfate- **SLES)**. It exhibits many of the same characteristics and is a higher foaming variation of SLS.
  - Contains low levels of 1.4-dioxaine classified as a carcinogen by the EPA.

Wiki Ref: **http://en.wikipedia.org/wiki/Sodium_laureth_sulfate**

Ammonium Laurel Sulfate (ALS) (ammonium dodecyl sulfate or CH3(CH2)10CH2OSO3NH4).

Is another surfactant that in high concentrations is an irritant to eyes and skin. Shampoos containing up to 31% ALS registered six health complaints out of 6.8 million units sold. These complaints included two of scalp itch, two allergic reactions, one hair.

damage and one complaint of eye irritation. So, ALS gets no complaints when used in brief discontinuous applications not exceeding 1% concentration. But complaints are not the major concern. It is the cumulative effects that we should be watching for.

Read Wiki: **http://en.wikipedia.org/wiki/Ammonium_lauryl_sulfate**

Parabens: a group of organic co-polyesters generally used to inhibit the growth of bacteria and fungal growth. Below are several that would never be tolerated in any quantity around animals) Used commonly in all cosmetics, moisturizers, and "so called" antiaging products and as a food flavoring since it is antimicrobial and antifungal. Even though it is an obvious poison, it is allowed in concentration of up to .8% in a mixture and .4% as a single paraben. All cause chromosomal aberration in animals and should be avoided in any concentration. The human body can absorb as much as five pounds of cosmetic chemicals every year. Parabens can mimic hormones in the body and disrupt functions of the endocrine system The acid forms show up in most commercial shampoos (and they are strongly acidic even if in low proportions) .

When Parabens are applied to skin, studies show that 60% of it enters the blood stream... and almost no commercially available cosmetics are paraben free per this website: **http://www.buzzle.com/articles/paraben-free-cosmetics.html** There are bound to be ramifications down the road when using them. You will pay the piper, eventually.

Read more: PubMed Ref: **http://www.ncbi.nlm.nih.gov/pubmed/19101832**
Ecomall: **http://www.ecomall.com/greenshopping/cosmetic.htm**

- Butylparaben (**BP**) C4H9(C6H4(OHICOO) is thought to do this by inhibiting DNA and RNA synthesis. So this chemical is anti-aging because it kills you... A new meaning!
- It was once used as an insecticide and rodenticide till it was **made illegal** in 1998 by the **EPA**. The median lethal dose in mice is 5gm/kg body weight. So it takes very little to kill a mouse. However, it is still allowed by our government in face creams, etc.
- All Parabens have been identified as "endocrine disruptors," meaning that they interfere with hormone production, balance, and regulation. So it shows up as: early puberty, breast cancer, low testosterone in men and unbalanced hormones women.

Wiki Ref: **http://en.wikipedia.org/wiki/Butylparaben**

- Benzyl-parahydroxybenzoic acid (**BHB**) is Ketone of Cholorophenyl

- Butyl-parahydroxybenzoic acid, a carbolic acid copolymer preservative

http://www.chemicalland21.com/lifescience/phar/p-HYDROXYBENZOIC%20ACID.htm

- Ethyl-parahydroxybenzoic acid (**EPB**) C6H4(OH)COOH·2H2O are colorless crystals melting at 210°C; soluble in alcohol, water, and ether; used as a chemical intermediate and for synthetic drugs... Another reason not to take drugs!
- Ethylparaben (**EP**) (HO-C6H4-CO-O-CH2CH3) Same uses as the Butyl form above it is a B-Glucan inhibitor that attacks cell wall membranes and stops mitosis... very scary!
- Methylparaben (**MPB**) CH3(C6H4(OH)COO). is a preservative, bactericide, fungicide, and has been shown to build up in breast tissue. It is commonly used in cosmetic creams and foods. It is easily absorbed through the skin. It increases the effects of the sun's UVB damage and, thus, increases aging effects and DNA damage. Are you getting this!

WIKI Ref: http://en.wikipedia.org/wiki/Methylparaben

Methyl-parahydroxybenzoic acid (**MPB A**)

http://www.chemexper.net/ME/methyl_p-hydroxybenzoate.html

- Propyl-parahydroxybenzoic acid is an irritant and acid

More obscure products used in commercial skin care products and shampoos:

- Dioxins: "dioxin" refers generally to compounds which have a dioxin core skeletal structure with substituent molecular groups attached to it. For example, dibenzo-1,4-dioxin is a compound whose structure consists of two benzo- groups fused onto a 1,4-dioxin ring.

  Dioxins commonly denotes the chlorinated derivatives of dibenzo-1,4-dioxin, more precisely the polychlorinated dibenzodioxins (PCDDs), among which 2,3,7,8-tetrachlorodibenzodioxin (TCDD), a tetrachlorinated derivative, is the best known. The polychlorinated dibenzodioxins, which can also be classified in the family of halogenated organic compounds, have been shown to bioaccumulate in humans and wildlife due to their lipophilic properties, and are known teratogens (cause birth defects), mutagens (cause mutations), and carcinogens, but all dioxins should be avoided like the plague... And *every* scientist knows this. They are just bad news.

- Diethanolamine (DEA) a wetting agent in shampoos, lotions, creams and other cosmetics. DEA is used to provide a rich lather in shampoos and keep a favorable consistency in lotions and creams. DEA can react with other ingredients like SLS in a cosmetic formula forming the extremely potent carcinogen, NDEA (nitrosodiethanolamine). NDEA is readily absorbed through the skin and has been linked with stomach, esophagus, liver and bladder cancer for good reason.

- MSG monosodium glutamate, yes, the food flavor enhancer, is not a good idea there. However, no one knows how it works as a cosmetic, so when it is used for that, you are basically doing their testing for them. One thing that we do know is that it is a salt of glutamic acid. It may be fine, but who knows?

- Propylene Glycol ($C_3H_8O_2$) makes a great organic alcohol antifreeze for cars compared to ethylene glycol, which is a poison, but you are a fool to be around this stuff for any length of time even if it is non-poisonous in small quantities... and it is commonly used in cosmetic products. It is a petroleum product converted from propylene oxide, using a catalyst generally, but not always.

- Sulfuric acid (H_2SO_4).... Come on now! What more needs to be said! Stay away from any product containing organic versions of this chemical... Like SLS!

http://www.livestrong.com/article/150690-what-are-the-dangers-of-parabens-in-skin-care/
Link to several effective homemade cosmetic products that work including:
Facial Mask, Cleanser, Toner, and Lip Gloss... So you can make your own and be safe.
http://www.buzzle.com/articles/homemade-cosmetics-recipes.html These are simple too.

48. Vibrations and Harmonics (4th edition)

Below is an important Dr. Mercola 2/17 interview with a cancer survivor who has written her book as discussed here.

*http://articles.mercola.com/sites/articles/archive/2017/02/12/cancer-build-your-own-cure.aspx?utm_source=dnl&utm_medium=email&utm_content=art1&utm_campaign=2017

0212Z1&et_cid=DM133741&et_rid=1883159839

Some key points touched on as I move on to the topic at hand are:

First, the conspiracy/ fact that all oncologists are on the dole with
drug companies (per Mercola):

Cancer drugs are by far the most expensive of all drugs (and likely the least effective against what they treat). Oncologists all sell these drugs and receive about a 50% kickback from all sales and this is unique to the profession. So as Mercola contends, there is virtually no incentive to make any cancer treatment a cure (But no cancer drug actually cures cancer, even the so-called genetically based drugs). For life threatening cancers, the best these can do is extend your life maybe two years (and I have 1st-hand experience here with my brother who bought into this lie).

If an oncologist wanted to improve a set protocol, thus, tried to be innovative, for instance, if he told you to do something outside the prescribed system, he would be reprimanded and finally could be sued for malpractice. This being the case, why would you even consult an oncologist other than for a diagnosis?

They do not hit on this, but Mercola discusses it elsewhere: The common FDA and MD recommendation today is a low, unsaturated fat diet. This literally makes cancer more aggressive if you already have it and increases your chance of getting it. That is, moving into the cancerous state or (vibration as I discuss below and as I discuss in elsewhere in this book, if you have not already contracted it.

Further Mercola discussions in the above interview:

Creative Medicine: He mentions a case where a South African physician recommended a high fat diet and was sued. However, today, you can hear several physicians online who have discovered that saturated fats help evade cancer and especially monosaturates.

Sun heals: He discussed how sun exposure leads to wellness and the implications.

Epigenetics: How your attitude and spiritual wellness protect you from disease. Thus they raise your ability to fight disease and thereby create a cancer cure.

Statistics: When the woman interviewed contracted cancer, the rate was one person in six who actually got cancer. Now the rate is 1:3 for women and 1:2 for men as the rate climbs toward 1:2 for both sexes (per Mercola) and your government is helping raise this number.

Healing Supplements: Some mentioned discourage cancer and also help heal it naturally by putting your body in tune (as I discuss below):

Cannabis: How THCA, the bio synthetic component cures disease, Read:
*https://en.wikipedia.org/wiki/Tetrahydrocannabinolic_acid *
for the pharmacological effects of THCA.

Burburine: A plant alkaloid revered in traditional Chinese and Ayurvedic medicine, but largely ignored elsewhere, cures disease. It is poised to become one of our most powerful natural therapies for preventing and treating a wide range of conditions and induces AMPK activation:
http://www.sciencedirect.com/science/article/pii/S0006295211008252

Mercola ends this interview by saying that cancer stems from mitochondrial health as my book contends and I further build on below.

So what did Mercola miss? First off Burburine is a good supplement if you are working with Mercola's supplement budget. However, for the rest of us, HL MSM is nearly three times as effective at doing what he is looking for when combined with the inexpensive supplements for almost all, as outlined in my book.

For unknown reasons, Mercola has missed out on MSM's effectiveness, even though he has interviewed suppliers and several experts on organic sulfur often. However, if you can find a supplier of burburine for $40 a pound, you have a great anti-aging and health supplement that rivals raw cacao, another of my favorites (at 1/3 the price).

How it Really Works/ Frequencies, Vibrations, and Harmonics

Your body emits a harmonic sound frequency and a resulting light emanates from that sound. Some call the light an aura and it can actually be seen. If you are highly in tune, this aura can fill up a room as many have reported with religious leaders and others. Thus, If you listen carefully, and are attentive, you can hear your own sound in your ears. You have no less potential than any leader. Never underestimate yourself.

Any intervention such as Burburine (mentioned above), for instance, is a frequency or series of them if it is complex set of harmonics. The sum total of Burburine's frequencies is thus, a disease countering frequency (for most people). A cancerous person emits a far lower frequency than a vital person with no disease. All foods, vitamins, and minerals are also frequencies, even those that make you ill are frequencies. Some may not be in tune with your body, so Burburine may be excellent for most, but counter to a few and this is true for all supplements and foods.

All disease organisms produce frequencies also. These will counter wellness in some people, but, of course, promote their own wellness in themselves. All organisms need to survive as nature intends, just as you do and, thus, if they counter your own wellness and optimal frequencies, can make you ill if your immunity does not repel them. Other organisms are beneficial and totally essential to our survival as science is learning today.

Also, adding to the above, our mental and spiritual outlook affects all of this according to our condition. That is, we can raise our vibratory rates by just altering our attention... that is, thinking positive thoughts. So allowing our minds to drift (generally a good thing) actually affects physical outcomes and our actual biology in diverse ways just as some have been reporting for years, but herein is the mechanism and it is not so difficult to grasp.

What we find is that any infection can be cured with a proper intervening counter frequency (or antibody in this context). The body cures itself with antibodies, which can be physical entities that we make to affect the negative condition that is causing the illness. Thus, given enough time, the body can produce its own antibodies and heal itself. The goal here is to balance all vibratory elements and give them added strength by raising mitochondrial counts and their vibratory rates. This can go even further. Communities of people when stressed from an antagonist can produce antibodies as a group, called the herd affect in science. This has been proven historically.

The mitochondria are the energy producers in each cell. They commonly vary in number between 50 and 2000 average per cell in people. Each one and all cells together contribute to your overall frequency. Therefore, as Mercola understands on some level, even though he may not be familiar with this entire mechanism: A higher mitochondrial count per cell helps your orchestra play with strength. When it plays in tune (frequency) and with high numbers (strength), it is virtually undefeatable by any pathogen or poison.

Adding to the above, water is the universal carrier of all of these frequencies and is thus the key to all of the biology on earth. So, a maximized, tuneful water is essential. This will carry these harmonics without distortion, keep them clear, and is easily understood across the band by the body receiving them. High water quality is essential to wellness both in people and also for robust plants as we have proven. Again, read, "Pretty Fine Sex is Spiritual" for more on this.

49. The Health Conspiracy (4th edition)

The Drug Culture and Arthritis (fun)

It is interesting that people actually buy into this giant scam along with the FDA who is we are led to believe, has our interest at heart. Below, I basically repeat an ad for the "#1 Rheumatoid Arthritic" drug, a prescription biologic medicine that I code name "**Embroyol**" here, with the real facts in parenthesis after their advertisement warnings: This is a literally lifted word for word TV ad with my commentary.

- **Embroyol "may lower your ability to fight other disease"** *(in fact, it must do just that, since , of course, it lowers your autoimmune response. this drug literally robs Peter to pay Paul).*
- **"If you have any other infection like the flu, do not start "Embroyol"** *(Why not? This, because lowering your autoimmunity level during any lowered immune condition can easily lead to death. So if you start Embroyol and unknowingly have flu onset, it could kill you).*

- **Do not use "Embroyol" if you have been recently injured or have cuts or bruises, etc.** (*Why not? You need to repair those and this drug will stop all repairs as it robs Peter to Pay Paul just as most do).*

- **Embroyol "lowers your ability to fight infections. It can lead to Cancer, Heart Disease," etc.** *(the list goes on, but basically, any other health condition including what Embroyol is designed to repair, can easily occur, since your immune system is compromised as with most drugs).*

- **Embroyol "can cause serious and even fatal health events if you have had heart disease or any serious related condition"** *(now lets be straight here: heart disease is the more serious of these two conditions, so do not try to rid yourself of rheumatoid arthritis if you have suffered from a heart attack using this crazy drug treatment, LOL).*

Moving on from the above Ad:

Your intelligent alternative treatment and the effects:

Interestingly, virtually everyone who takes the AHL gains full wellness within a year or so, even with rheumatoid arthritis, this writer included. By the way, rheumatoid arthritis is still considered incurable by most doctors today despite the advent of Embroyol.

When I wrote the first edition of this book, there was much to learn and virtually no one knew about high levels of MSM, much less the AHL. In fact, there are still plenty of unknowns, but the basic facts are now well in hand with the help of reader responses. By combining HL MSM with Whey, organic C, Sunshine (or vitamin D3 in winter), organic Magnesium, Resveratrol, Astaxanthin, Saturated Fats and CCO, and others that suite your body, you have the AHL and you simply get back to the wellness that nature intended. This book outlines those important vitamins, minerals, and foods, but be creative and do your homework. More is learned every day.

So should you begin this protocol if you have flu, cancer, heart disease, or any other serious condition? By all means, but why? It raises your immunity and ability to fight all problems. Generally, no one ever dies from an overactive body defense system, too many mitochondria, or too much energy being created by them.

The actions of MSM in combination with the above supplements combined with a low sugar, low carb diets are not so easy to grasp by studying the three most complex natural cycles, the methylation cycles that fuel the cells at the quantum level and slightly above. But this the key to the magic that occurs and it is not all about sulfur (as everyone reports). However, you need not understand what is going on there if you simply keep in mind that your body will do all it can to stay well if you give it a fighting chance. You are a survivor. Nature gave you a complex tool kit to beat all disease with the most important aspects beginning at the microcellular level where these cycles are always at work. Combine them with full hydration levels and you have the key to life.

What the drug companies are doing is trying to rob from one section of your biological tool kit and give that energy to another. That is Embroyol... It steals energy from your autoimmune system just as the drug company boldly announces. That theft can and will kill you if your system is compromised in any way as they warn. They mention flu in their ad, but even a bad cold could start this downward spiral and virtually any health condition that saps your microcellular energy qualifies as a killer.

Short term, your autoimmune system needs to put all of your resources to work curing you of the condition it recognizes. Note that the "Embroyol" ad even touches on cuts and physical damage as a reason not to use their drug... Imagine that! Your system is busy healing you of a cut and you throw a product at it that stops you from healing and turns your defenses in a direction that your body knows is not where it should be going. So, in effect, this drug company is put in charge of your immune system and is claiming superior knowledge. So tell me, how does any of what they are pushing make any sense and why would anyone start this drug in the first place?

No one really knows just what their body is working on at any given time. You may be very much in tune with your physical body and very aware of your energy levels at any given moment, but you may attribute your lowered energy level to your arthritic condition and it could be the beginning stage of a cancerous lesion caused by a toxin that you were exposed to while driving past a farm on the way to work. The result? You set yourself up for, as the drug company suggests, death from a serious liver overload and later possibly incurable cancer.

I am not picking on this "Embroyol" drug here. There are literally thousands of drugs and they all basically function the same way with very few exceptions. All drugs are designed to work against a single problem at the expense of total wellness and, of them all, cancer drugs are the worst as they, more often than not, end up lowering the immune system so far as to kill or cause the cancer to reoccur.

Finally, keep in mind also that this condition, arthritis, in any of its forms, is a generally a precursor to heart disease and this begins, not at the large arteries, but at the small capillaries that mainstream medicine has no answer for. The first really bad symptom, in my case, was that my arms were going too sleep at age thirty for no reason. I had had rheumatoid (juvenile) arthritis since age four (my first real awareness of its onslaught), but probably beginning at birth and my mother was undoubtedly deficient in organic sulfur during her pregnancy. The symptoms at age four? A deep and reoccurring aching in my right hip (I thought it was polio) and fingers that swelled and itched when cold (no playing in the snow for me).

I suggest that virtually every person born in today's United States is born with dietary deficiencies of some degree. My own mother grew up on a farm and ate well, but somehow she missed on organic sulfur and that is not so difficult to do since that farm never moved. Today, we know that within three growing seasons the soil can be depleted of sulfur. There was a time that people moved on, when the soils were depleted of minerals, and started over with a new farm. This is never the case today, but I can tell you that my mother was far healthier than the average reader here at age twenty-two when she gave birth to me. She ate all-natural fats (lard and animal fats), virtually no sugar, and my grandfather fertilized their vegetable garden with manure. That was a start, but it was still inadequate in my case.

The Plot Thickens (4th Edition)

The word "conspiracy" became popular with the death of JFK and is said to have been popularized by the CIA to offset the dispute of the "single bullet theory." In the context of health, it is never used, but I submit, that this is a very broad and across the board conspiracy as it about earning vast sums of money at the public's expense without their knowledge. Below, I expand on this:

The human body is no different than any other biology. It is well adapted to conditions and very good at survival. Note that only human beings believe that they must monitor their chemistry and alter it with drugs in order to stay well. Some, for instance, monitor their blood pressure on a daily basis. We are told that doing this can extend lives. However, when you discount deaths at birth and childhood deaths, the evidence is that we are not living as long as we did prior to those measures and those who never take drugs have lower death rates than those who do. For more on this, watch this talk by Dr. John Bergman: https://www.youtube.com/watch?v=p3V3TITSDxc This is both a humorous and serious talk at once. Watch it and have fun!

Also, note that the three greatest killers today: heart disease, cancer, and Type II diabetes are all increasing at alarming rates, but it is very difficult to quantify them, as the CDC does not publish just how steep these curves are in many areas: http://countryeconomy.com/demography/life-expectancy/usa Here it would appear that rates are decreasing, however, indications are that this is just not the case as Bergman reports above and, in fact, one in two people have life threatening diseases.

In fact, cancer deaths, as Mercola reports, have recently displaced heart disease as the number one cause of death, but all three are going up steadily in the US. Interestingly, this is the case when one separates out the data as Mercola has. Further, this only began to occur some sixty years ago when special interest groups began controlling the Federal Food and Drug Administration (FDA) and other three letter US authorities to the point that the controlled our markets and our environments. In doing this, they placed the attention on issues such as "global warming" that no one can actually control and took our attention off areas such as GMO crops which are entirely within our ability to correct. Its about getting our attention where they want it, where money can be made by corporate interests that pay taxes and make contributions to campaigns.

The Conspiracy Apparently Started with Heart Disease

This all appears to have started with the invention of margarine (from cotton seed) and the "saturated fat" lie which generally still prevails in mainstream medicine. If you doubt the value of saturated fats and especially butter and whole dairy read: **https://authoritynutrition.com/butter-vs-margarine/** as the fallout from this conspiracy continues to increase the rates of all three of these diseases and many less prevalent ones such as MS, Lyme Disease, and others. Or simply Google saturated fats and cholesterol and variations. This lie will go away eventually, but will you outlive it? This part of the conspiracy started with Ancel Keys as discussed earlier, but the food industries quickly profited from his ignorant premise and made unsaturated fats, vegetable oils and cotton waste a part of our eating habits. Today, the drug industry has joined in with drugs designed to fix this perceived notion as everyone is told to manage their blood cholesterol levels. This multi-billion (even trillion) dollar industry, then, simply adds to the problem.

Cancer, the Greatest of all Health Conspiracies

A Simple Study Using CDC (government) Statistics:

With cancer, let's pick one type, breast cancer. The CDC rates generally focus on locations, However, all cancer rates are generally increasing as reported at the 2016 "Truth About Cancer" forum. This, three day forum, was produced by Todd Ballinger with some thirty imminent speakers including Dr. Mercola, Mike Adams, and others. One premise often overlooked as with all of these killers is that while it most certainly matters where you live: it is a function of food quality, chemicals in your food, environmental quality, the drugs that you take, and your eating habits. To watch a repeat, go to: **https://go.thetruthaboutcancer.com/?a_aid=550894055d93f&chan=GCancerVideo&gclid=CPDJy6iiitACFYODaQodYCECGg**

My commentary on location: Looking at breast cancer death rates as a percentage of population: The rates in various countries vary wildly. The rate in most developed countries is generally around 20%, while the rates in the lowest five, occurring in less developed countries, is around 5%(?). Imagine that, four times lower rates. **http://www.worldlifeexpectancy.com/cause-of-death/breast-cancer/by-country/**

It stands to reason that the doctors in Bhutan, where the reported rate is 2%, are not ten times as good as those in the US. One might explain the difference with reporting, with these lower rates, but ten times or five times lower in Guatemala? These are all about life style and nutritionally related diseases and the CDC proves it with their statistics.

So disregarding the ten times lower rates in Bhutan. If our US cancer industry could report even a 50% lower death here in one year, it would be heralded as an outrageous medical breakthrough. So what is the differences that separate Bhutan from the US diets?

- First, no women likely get breast x-rays in Bhutan.
- Second, few ever eat sugar, being in a survival mode for the most part.
- Third, at the moment, no one there eats GMO foods there or in any undeveloped country.
- Forth, everyone there eats large amounts of natural saturated and unprocessed fatty foods. Now before you go running to your map, Bhutan is just south of Tibet and east of Nepal. China's rate is 6% and Nepal's 8%. So both are well below the US rates still as are all undeveloped counties.

Now a little math and small amount a savvy will quickly tell you what a death rate of 20% in the US means. A woman in Bhutan who does contract breast cancer does quite well and with little or no health care. Most women there probably go untreated, but who knows? That would be an interesting study, but what are the chances that anyone will do it? It is counter to US medical interests to do any such studies.

So, the point here is that the US is helping lead the charge toward sickness. Now looking at the US, with heart disease as reported by the CDC: In Nevada, Colorado, Wyoming and New Mexico, the rates are 50% lower than Mississippi and Louisiana. Mind you, this is the United States here, not third world countries. We can assume that they are mostly alike with all diseases other than such things as malaria and tropical infections.

50. The French Paradox Revisited

Actual numbers are very difficult to correlate. However, when discussing the French "paradox," seldom are actual rates stated and often the effect is reduced to wine, which is full of sugar (and is a huge industry if you have noticed). But alcohol is a poison, period. If you need proof, drink a quart of 100 proof. Too much and you die, but it is very popular with everyone, a huge industry, and reports across the board are biased as one would expect. If there is one thing that you can depend on, it is that any entrenched industry will generally make a good showing when put to the test. Wine, beer, and hard liquors are no exception here. These are now established drugs that our population loves and cherishes and the government no longer holds biases against. Certainly, hemp, now known as marijuana, is a new one that is beginning to make inroads and it even has some promising medical uses. It could easily become as successful as any alcohol and actually has some real, not falsely reported, health benefits. There is no way that drinking sugar and alcohol in any combination has any true health benefits, but you will keep hearing that it does.

Not to belittle the effects of Resveratrol, but have you noticed that real French cuisine includes huge amounts of real fats including butter and also, real bone broth? These are real foods from animals, real spices, and fresh vegetables, with nothing from cans. Those alone compared to our, "Can you believe its not Butter?" foods will surely make you much more healthy and it these effects will negate a small amount of wine in the process.

For information on why the CDC may not be giving us all that we need (or expect), read the book "Vaccine Whistle blower." As an aside, and I cover this in my books as related below, cancer is a microcellular based disease. It starts with environmental toxins, but moves on. I first was alerted to this when a reader of my first book, who had terminal mesothelioma, mysteriously became well when taking HL MSM. He was given three weeks to live. Once you look into the methylation pathways (cellular energy production) and relate them to MSM, its all ties together. More below on arthritis below, but as with all disease, diet is incredibly important and his wife reported that he combined it all.

Type II Diabetes

The rate of increase in diagnosed Type II Diabetes in the US, per the CDC, is approximately 1.6% per year and is now about 30% for those over age 65. By ethnicity, the rates in whites are half that of American Indian Natives. You might ask why?

The Chinese rate is 4.4% while the American Indian Native's rate is 16% or 275%. The disparity in is thus about 275% between them and the people of Bhutan and China are very close. Still, if you want to stay free of Type II, you do not want to grow old in the US and certainly not live on a US Indian Reservation, but they do not fare well with any of the big three diseases since the government supplies them much of their food.

The above could easily become expanded to a much broader sample, but I submit that the results will be the same. The conclusion that one can objectively draw from the CDC data is that we are being systematically killed by the food and drug industries and, from these references, and this is being backed by related government agency reports. As Bergman will tell you, cats do not have health insurance, nor do giraffes and, by the way, there are no giraffe MD's. We are finding ways to kill off our human population and at an alarming rate as time goes on. As many, including Dr. Joseph Mercola will tell you on his website, **http://www.drmercola.com/articles/**

From this, it is clear that you do not want your government supplying you food and school age kids are becoming victims of our greed as they consume high levels of sugar, carbs, and unsaturated oils in school. Today, there are 5000 new school age cases per year and the rate is increasing at 5% per year. So the term should now be Government Induced or School Onset, not Adult Onset Diabetes and the disturbing effects include blindness and heart disease.

As of yet on Type II does not clearly indicate how the methylation pathways and HL MSM affect its progress. However, the previous term, sugar diabetes, was more accurate than most know today. Certainly, the AHP will allow your body to overcome most of these complications and allow it to heal as Type II is entirely dietary in nature (and our government backs the cause). No drug can reverse a poor diet, ever, but for adults, there are drugs for Type II (?) that are not approved for children, so in that way, kids are lucky.

Arthritis

This has been discussed many times here and few ever consider any type as curable, but there are basically two types of arthritis as outlined below. Moreover, they are absolutely connected to coronary artery disease, the number one cause of death here in the US, heart attacks. but current science has never made this connection. They have seen the relationship with Type II diabetes, but the three are interrelated diseases based on low fats and sulfur availability and microcellular energy rates as discussed elsewhere herein.

First, all drugs currently sold to help with Arthritis are literally poisons and as this article says: **https://w3.brownsteinhealth.com/Health/DRB/LP/Brownstein-Arthritis-Proof-Trial?dkt_nbr=sjn3hnoj** "There is no Cure for Arthritis." While this is incorrect, certainly, no drug will ever help with arthritis of any kind and all joint disease can be handily cured within a year and most can be cured within three months as reported here and elsewhere. Still it does not end there. Read the new posts at the end of this book that tell us that this HL MSM protocol has provided in feedback more recently. None of these extended are befits, especially cancer were anticipated when I started this book and they are appearing across the board along with others like prostate problems that were never anticipated at all.

With all of the above, if you plan on living very long, statistically, stay out of hospitals. The statistical death rates from people passing through them are staggering (trauma care aside). Mike Adams, the owner of a health testing lab and contributor to many current web videos and author of many books will concur as follows:
https://search.yahoo.com/yhs/search?hspart=mozilla&hsimp=yhs-006&ei=utf-8&fr=ytff1-yff28&p=the%20health%20ranger%20mike%20adams&type=

51. The Methylation Cycles - MSM's Secret (4th Edition)

Herein, I discusses how the Methylation Pathways are acted on by MSM/ DMSO and how important this is to your health. The chemistry is that the two methyl (CH3) groups and the (S) sulfur. DMSO adds the oxygen group and is thus the oxidative state. Therefore, the two methyl groups comprise two thirds of the MSM molecule and are totally ignored by everyone today as previously discussed. This is to say that when you have adequate stores of organic sulfur, unlike what is commonly assumed, MSM/DMSO still has a demonstrable effect on your health and wellness. Therefore, HL MSM must contribute to your wellness and de-age you with rare exceptions.

Much has been written about MSM/ DMSO since I started writing this book some years ago. In those days we had a couple of pioneers, but no one then understood the significance of the methylation cycles. Still today, virtually everyone, attributes MSM's amazing health benefits to organic sulfur alone, even though a few get the high level part to some degree.

Sulfur is indeed a key component and it also does contribute to one of the three methylation pathways. However, even though many scientists and healthcare professionals are well versed in the many health benefits of MSM and a few have begun to understand DMSO as an oxidizer, no one has associated MSM/ DMSO to methylation, the key to life itself, as everyone agree it is.

The Methylation Pathways are the most complex cycles in all of biological science and are also the most important to any and all outcomes including sustaining optimum Liver Function as well as optimum:

- **DNA and RNA**
- **Wellness**
- **Detoxification of Chemicals**
- **Immunity to Toxins**
- **Weight**
- **Energy**
- **Cellular Repair and Replacement**
- **Stress Reduction**
- **Mental Wellness**
- **Hormone Processing and Production**
- **Enzyme Production**
- **Antiaging**

The above are a result of several key functions, but increased glutathione production, the master antioxidant, which affects MTHR, will always stand out the most significant contributor here. Virtually all disease is due to methylation resistance, but as long as you have adequate stores of glutathione and your mitochondria are at high levels, you will likely do very well.

The above is the bottom line here, but for those interested in doing their own investigation there are plenty available. Below is one good site for diagrams:

https://images.search.yahoo.com/yhs/search;_ylt=A0LEV7gShqRYyXYAFSQnnIlQ;_ylu=X3oDMTByMjB0aG5zBGNvbG8DYmYxBHBvcwMxBHZ0aWQDBHNlYwNzYw--?p=Methyaltion+Cycle+Diagrams&fr=ytff1-yff28&hspart=mozilla&hsimp=yhs-006

For more expert information on this, watch the YouTube interview with Chris Kresser, L.Ac, who in and expert in Methylation:

https://www.youtube.com/watch?v=MktZ4E8Ym54

Also, listen to Carolyn Ledowsky's discussion" How Can MTHFR affect your health:

Herein, she mentions shape changes in genes. At the microcellular/ viral level, these changes affect the shapes of the actual physical organism and, thus, how they affect a host. This causes genetic mutation resulting shape shifting mutations and problems with the resulting viral and bacterial diseases Lyme Disease, especially is one, but there are others that "escape" when high levels of antibiotics are employed to counter them.

https://www.youtube.com/watch?v=gC-NAfB0qE4

52. A Summary:

My diet today is not so much different from that of Dr. Wahls', but many people are just not ready to go there. Unlike them, I can generally learn to enjoy foods that tell my body to be healthy. Whey is one such food that has worked for me... MSM is another. It must be good if it keeps me this well, so yum.

On the other hand, I learned to love refined sugar as a kid, but now I see it as a slow poison. Soft drinks/sodas were essential to life in Air Force Boot Camp. They became more even important when I was not allowed access to them, but now I have no desire for them at all. I admit that I still love the smell of home-baked bread, but even that, I pass up, for the most part. Its all a learning experience and we learn, even at the cellular level.

Also, overcoming a life threatening disease like Dr.Wahls' MS is a huge spiritual asset... and we can certainly build on these assets. Dr. Wahls' contracted it and has thrived. It may have been terrible for her to endure at the time, but she is a greater person as a result.

If you get in the habit of eating the Low Carb Diet, Undenatured Whey Protein, and supplementing *HL* MSM on a daily basis, you too can accomplish much of what Dr.Wahls did out of the desire to survive. Would Dr. Wahls have done that without the fear of death hanging over her head? Only she can answer that. But below I outline why this book shows you the way to the things that Wahls could not access:

The Bottom Line

I have outlined many scientifically irrefutable methodologies herein, but they are all about cheating the procedures that we commonly accept in our diets and daily lives. They include:

First MSM/ DMSO: This oxygen transport pair that absolutely alters the three basic energy cycles at the microcellular level and cause the mitochondria to go into what I refer to as a "Supercharged" rate of energy exchange. From here, I add the following adjuncts:

Interval Training: an exercise program that allows all of these super supplements to flow through your system. Interestingly, this is not a particularly intense system. Tennis twice a week or any racquet sport qualifies, but this is key for joint health and joint health is the precursor of hearth disease, the killer of the masses who believe in what the FDA teaches.

Drugs: the FDA and big pharma dictate what you are told by the mainstream media and most professionals today. This is not information that will keep you well. Even the commonly published statistics will lead you astray as Dr.Mercola publishes on his website. Hopefully, Google will allow him to continue, but they shut down The Health Ranger,

Mike Adams' website in February of 2017. Expect more of this behavior as we confront them with facts and many of them that they generate out of necessity.

So what are drugs and why can they never work? This story is told elsewhere herein, but the bottom line is that pharmaceutical science promises an incomplete science as follows and it is based on the ownership of patents: Scientists look for natural nutrients that have been identified to provide some relief from disease. These may be in the form of a natural herb or even a common food. Next, they attempt to isolate the active ingredient for patent and sales, thus magnify the results and cure a disease or provide disease protection.

There are two serious problems that keep the above from ever proving out, long term, and, always, there are serious consequences as I point out herein. Please allow me to point out two again: The first is that when isolated, these the herbs and foods lose many of the active components that help your body to cope with them. Next, while the companies are required to pass rigorous tests individually, it is totally impossible to test two or more of these identified factors (drugs) against each other. This means that when you are taking two or more drugs, no one knows scientifically what the interactions will be. Today, many people, believing in the FDA and their doctors (often several specialists who are unaware of each other's advice), take ten or more of these radical drugs. Without knowing your combination, one can guarantee a good outcome and this is not good science no matter what you are told.

From here, I have offered many ways to cheat the hangman, but the most powerful is spiritually. There are thousands of spiritual people on earth that you will never hear about who live for hundreds of years. They teach a few their secrets, but they are not allowed to makes them common knowledge. As masters, they know all and must follow the universal rules of spirit. I know a few and have direct knowledge of their craft. So a basic spiritual rule is to never give out this information to the masses as it is simply too disruptive. Still, if you are willing to invest the time, herein I give you some key "tricks." Go back over what I reveal herein and invest your time (love). So, adding to the basic MSM/DMSO trick:

- The key to all of this in terms of nutrition is to eat the most simple life form before any of our "advanced" technologies ruin it as follows:

- Note that the largest creatures on earth, the baleen whales http://marinelife.about.com/od/cetaceans/tp/types-of-baleen-whales.htm eat the most simple forms of food, phytoplankton https://en.wikipedia.org/wiki/Plankton. our interest here is in krill as it is a very clean and unaltered food source that you can purchase and supplement (Astaxanthin as sold commercially).

- Next up the food chain is very small ocean fish which we commercially call sardines. Sardines are not a genius or species of fish, but are a clean and generally unaltered form of food that you should learn to eat if you have not.

- Again, the idea is to enjoy foods that our technologies and farming methods have yet to have ruined. In walking through your food store, you will find maybe 1% (I am being generous here) of what they sell is really food and therefore, you must walk past it if you plan to live a long and healthy life.

How to Cheat

Aging Tricks: One of the key ingredients in this process is antiaging. In fact, you must age backwards and this is key to wellness. Herein, I tell you how. If you missed it, go back over those sections in greater detail. If you are over the age of fourteen or thereabouts at the cellular level you are prone to disease.

The masters who maintain youth and live hundreds of years have mitochondria counts typically at 2000 per cell average coupled with outrageous mitochondrial efficiencies. You must get your cellular age down to this range. Once you do, you avoid all disease. Do I know that this is possible?

I do not have proven count at these levels, but I stopped ever getting disease over five years ago now. The most common disease is a cold. No master or their students ever catches a cold. For those who want to forgo antiaging and just be well, there is no answer. They are one and the same. That is, you get one when you get the other. Also, you are not going to get either unless you begin the path to spirituality which is about love and personal responsibility

Supplement Tricks: There are little known advanced food forms beyond MSM/DMSO as follows that modern science has devised:

Liposomal Supplements: Start with Lipo C that you can make yourself, but there are advanced transdermal melatonin and others... https://www.absorbyourhealth.com/what-are-liposomal-supplements/
http://www.livonlabs.com/cgi-bin/start.cgi/liposome-encapsulated/high-quality-liposomal-supplements.html
We must have lab proven encapsulations and not just emulsions as the above state, but given you have this, you are in a new world of supplementation that was previously unavailable.

These supplements are cutting edge and easy to digest. In fact, many are transdermal and avoid the need for digestion totally. However, they are easily five times the cost of normal supplements, but infinitely better. The main risk with these is that they do not mix well with drugs. However, they likely mix far better than any two pharmaceutical drugs could. Chapter 30 herein discusses this huge problem among others. If you can fit the liposomals into your budget, you will be amazed at what they can do. These are Lab made except as above, but this is one place that you can enjoy the benefits of science and not be concerned. Also, the transdermal varieties are especially simple and effective .

In discussing liposomes, one would be remiss not to mention Alpha Lipoic Acid (ALA) which has been marketed for many years by health food distributors. This organic acid is used in the compounding of many of these products, but ALA alone has many amazing tributes in itself per https://en.wikipedia.org/wiki/Lipoic_aci and it is utilized in Methylation cycles

https://www.nanonutrausa.com/product/liposomal-vitamin-c/

Colloidal Minerals: Minerals should be your primary center of attention, since all that vitamins generally do is help you absorb minerals (Vitamin D is not a vitamin), as Wallach told us thirty years ago and I repeated earlier. Colloidal particles are in suspension and are between 1 and 1000nm. We want the smallest particles possible and minerals that never settle out. https://en.wikipedia.org/wiki/Colloid . The trick here is to avoid the need for digestion as much as possible as with liposomal supplements. But these, unlike liposomes can be from natural sources and can be inexpensive to supplement.

If you have been diagnosed with cancer, this is your best chance at moving back to health and many have already proven that it can be done. Nevertheless, no one should be working from this point of view. You need not wait for such an event to occur. Cancer is always easier to avoid than to cure and optimum wellness precludes the cancer state. So there are proven clues here to help you do that, but it is always up to you to fill in some of the blanks. Each of us are all a little different. Our recipes vary. It is our personal journey.

And just maybe, we can all experience growing 800 years old as this small group of individuals who were "listening" carefully to the feedback that we were getting as individuals... and not following the heath advice of the mainstream "experts" (who do not appear to be faring that well as a group if you have noticed). When we can, "just get'ter done" in one shot . . . if that is what it takes.

Optimal heath is a closely-held secret and, most certainly, it is one held by far less than 1% of the population. But despite this, it actually occurs. However, it is in fact dismissed by most members of the alternative health community as an impossibility. In the next 50 years, we have a lot to prove. Dr. Mercola's site reports the statistical downside of the mainstream failure better than just about anyone... and it is obviously all true. I too was among the doubters and did not know the difference. Thank God for this world wide web communication network and these technologies that we have developed as our ideas and words are passed around with constantly growing results.

I am just a kid... Born in '42. There is plenty more to learn, but think I am up to it. We have much to thank the brilliant Dr. Seneff for and her contribution to this story (especially her last report in this book that seals it), but others could be coming out of the woodwork as it unfolds. Maybe: Dr. Mercola, Byron Richards, Jon Barron, and Mike Adams will finally hear my story and "get it." It's not that hard to grasp, but most importantly, it's true, it works and it has shown to be increasing repeatable by the readers of my books! Each of the above experts hits on key points already, but they all miss the final story. However, the real goal is to transform the way medicine is practiced, generally and they are doing a fine job at this... To stop the slash/ burn/ chemical destruction of so-called cancer therapy... To avoid most joint surgeries and most surgery interventions in general... To do away with all of the heart surgical interventions. All of this can generally be avoided with this Protocol and it is very inexpensive with no doctors needed.

What we are discussing, on a grand scale, is virtually a new way of living that includes a more open form of government with far less intervention and more creative small businesses. The result is less destruction with less effort and less energy that produces more effective work. In this vision, almost the entire $2.6 trillion Drug Industry would be unneeded along with the Health Care Insurance Industry and myriad of others required to support it. Think about it. Less than one hundred years ago, these did not exist.

What we are finally envisioning is a population that is at least three times more efficient and a whole lot wiser, and happier... with more of the free time and freedom necessary to dream up our next level of beingness. Then, only our fragile minds and our reliance on others can hold us back.

At the same time, however, we must be given free will... and the freedom to fail. Its always our choice as individuals and as a country. We are limitless beings in a world that is telling us to get on a linear train as we enter a tunnel with no real destination... with no vision of where it will lead except more corporate gains and government intervention.

As this train gains momentum and steam, everyone, especially, our leaders grow fearful of jumping off... they might get hurt! Some call it a rat race and that is currently an apt description, but we stick with the same leaders... who have no vision and no guts. To me, a crash may be the best option. At least then, we can work to assemble a new method of transportation and move on in directions that work.

Looking at the problem from the current narrow popular viewpoint, my vision is clearly impossible, currently. But we should, as future spiritual masters, always be viewing the whole from the widest vision that we can muster. Moreover, we, as spiritual beings, are blessed with the ability to actually pull this all off through epigenetics and beyond... with unlimited potential as we learn to Thrive.

This is the same vision that Thomas Jefferson and the founding fathers worked so hard to implement by bring it back around and functioning as they had hoped. They wrote our Constitution in an effort to protect us from the overpowering influence of a destructive and ever growing government, but we have allowed their effort to take a back seat to what they feared most.
Maybe some savvy and caring doctors reading this will get onboard and help us carry this story to the mainstream press and finally to the masses who are generally mostly interested in "being happy" and trying their best at being sexy... Not that I don't enjoy aspects of these also.

However, at a practical level, I can guarantee them that their deep transfat GMO oil fried, potato chips and fritos will not get them what they want nor will it sustain them on the train for long. It is a total dead end, but the dead end is an illusion. There are ways out and nature actually has provided them and at a significant discount.

As Mercola says (eyes wide),"Take Control of Your Health."

I reduce that down to, "Educate Yourself." Its probably best reduced to, "Pay Attention." Nothing comes for free, really, but maybe it can start right here.

One young woman in a Dr. Oz interview was reporting on her book that was touting raspberry isolates as a way to control weight. Dr. Oz asked her how she got the information and her reply was, research, research, and more research. It is true and we all have that ability now with the web. Thank God.

In 20 Years, How'd it Work?

In the above 1996 photo, I was smiling, but I had to wear a steel knee brace to play tennis, Sometimes, without warning, the knee would collapse and I would end up in a pile. My neck hurt constantly and I would have reoccurring tennis elbow. In cold weather, my fingers would seize up and I could not hold a wrench in my hand. I had blinding migraine head aches about every month and occasionally more often. These would generally keep me in bed with horrible pain for a day and they would not totally be gone for maybe two days. I had occasional chest pains that I shrugged off as normal. My friends reported the same pains, so I was OK with them. Also, my "arms would go to sleep" during the night or when lying on the couch. I was, in a word, Dying.

So what is the difference in about twenty years of aging? There are no identical twins here, but using your body as a primary test bed without scientific controls can work well IMHO.

Ideally, in my experiment, as Soul, I would control two bodies and give one our AHP while giving the other the SAD (Standard American Diet). After 20 years of comparing the two, I would have first-hand knowledge and an absolute comparison as I left the SAD diet body side with a smile. Maybe that will happen some day, but more likely, I contend that we generally all have more important jobs to do spiritually than just doing health experiments. We will see, though, just how this pans out.

In any event, I am stuck with just this one body to experiment with. As the saying goes, a picture is worth a thousand words, so here you go:

The 1996 picture was taken in 1994 in the Italian Alps near Lake Cuomo. As I noted above with the many joint problems, I wore a knee brace while playing tennis. I was slow, not too quick off the mark, and not nearly as agile as I am today and my game has improved a lot. I also had yet to suffer my first real sciatica attack, so I had plenty to endure in the next few years.

So, you be the judge. I started the *HL* MSM Protocol at age 56 in 1998 just after that picture, but lots of surprises were coming. By the 11/14/01 picture at age 59, I was wise to all of that part, but I still was getting migraines regularly. By 2014, they were very mild and rare as I said before. Now, they are gone, but I do get the aura occasionally.

By the most recent 4/29/12 picture at age seventy, I was well past all joint problems... A distant memory and today, I just never got colds or ill from anything.

I got the flu that year for one day and a fever of 102. My nose ran for two days as I had failed to change from my 5,5000iu/day fall mode to 10,000 iu/day winter mode. But I was listening, and immediately went to 50,000 iu of Vit D3 and took my last MegaHydrate tablets with the results that I have come to expect. Good hydration, as Dr. Flanagan will tell you, is a primary key to wellness and getting back in balance.

A failure to change and adapt is at the heart of just about any problem and this was a reminder that I had not adjusted. That was actually the first real illness since I was overwhelmed by salmonella from eating a Caesar Salad about ten years ago. Traced to infected eggs, at least this was proof that my favorite restaurant in those days was using real ingredients.

I did get an occasional cold till I discovered that colds are basically a deficiency in sunlight (Vit D3) as I relate elsewhere here. Read the Stephanie Seneff report at the just added to this book to see where this is going.

Suddenly, given the above observations, this book reveals the answers to most health conditions that the FDA is spending a large part of the budget on today to resolve. Stephanie tells you the problems and why and herein I give you simple solutions derived from what we have both observed in terms that all readers are reporting as working for them. You do not need huge amounts of government money or high technology to resolve them.

I had not started the full AHP until much later, in 2009, actually. So the above pictures are not really a fair test, but this *HL* MSM was the protocol that basically changed my life completely. Would I have aged less on the full AHP? The next twenty years will tell me more. Certainly, I still age, but now aging is fun and painless without any illnesses or loss of mobility.

I wish all of us the best of luck in minimizing the effects of cellular aging. I am pretty sure that we have the basic elements down well with the AHP, but improvements and new discoveries like Moringa, Cacao, Maca, and others will always evolve. There are thousands out there to discover, especially in the various mushrooms, oriental herbs, and sea plants. My premise here is that the answers are with us now and they will not come from the drug companies that our mainstream media reports on daily.

Then, adding the wonderful new Quantum innovations of Dan Nelson and his creative group that we learned a great deal from in our own pursuits.
So, be on the lookout and be willing to take the time to do the research. The web is still a marvelous resource when skillfully used despite recent changes. Some of these advances will be revealed in the referenced online discussions.

It is Better than I Reported Before

So what do I report in this at age 75 bottom line? First, none of the above health problems, ever: no flu, no colds, no migraines, no arthritis, no problems and my tennis game has impoved, lol. I have never been in this condition in my entire life. Readers are beginning to report similar results, but this will take time to bring out totally.

My conclusion here is that it actually takes maybe ten years to reap the full benefits of the the AHP, the HL MSM Protocol, and it is the higher energy levels from the methylation cycles as discussed more fully in my book, "Methylation, Awareness and You" that make this final phase possible. Also, Rich's water making device does not hurt, but that work is still ongoing. We are developing that device for increased plant growth levels currently. It is impossible to separate all of these results out into this single overall win, but this combination is a winner for sure.

This concludes the body of the book, "It's the Liver Stupid " with researched information.

Below are first the early on-line discussions that led to the above followed by new commentary, and the resulting book. The discussions at the 4th edition section ***were submissions by the author resulting from interesting reliable members and occasionally experts that pertain to the book as it was being edited for this version, combined with the author's commentary and on-line responses.***

53. Early Online Discussions History

Early online discussions are included as Historical background to show the reader how the AHP (evolution) and especially this HL MSM came about.

MSM Works for Emily
Hi Emily:

As to containers: It would appear that, especially if you have large amounts or do not use it fast enough, you should re-contain your MSM in a glass jar with a tight lid. The reason being that MSM picks up every chemical in its proximity ***(While many natural foods cleanse the body, MSM and DMSO may be the best of them at doing that anywhere)***. This is a very good thing once you ingest it, because it cleanses your body, but it could do the opposite if it has been sitting around for months in a plastic container. This is just a hypothesis, but it makes sense.

This goes along with why we recommend starting slowly on the protocol and why it works so well as a body cleanser. Of course, everyone knows this about DMSO, because it is so active. It would be totally foolish to keep DMSO in a plastic container even for a day *(IMHO)*. As most everyone here knows by now... MSM is just an oxidized form of DMSO, so the precautions still hold even if it is less active.

Finally, this is also why I recommend buying your MSM from a fast-moving supplier like one who sells it for horses... and the faster they move it, the better. As I have related here several times, my first knowledge of this wonderful stuff came from a guy who sold it by the rail car load and he shoveled it into a bucket for me. That was still the best deal in every way. The bottom line is that any handling makes for contamination and that is what you want to avoid.

Cheers, Huuman

MSM Works for Emily

Second Message posted from her:

>I had an interesting experience this week. My lab tests reveled, my thyroid was suppressed, so the Doctor cut my synthroid prescription in half. Now for the interesting part: for about 2 months I have been taking 3/4 TB. MSM every morning. I guess it works!!!!

I'll do lab tests again in 3 months to see how I'm doing. Predictions anyone????!!!!!<

Hi Emily:

I don't know anything about you and you reveal nothing except your basic problem. However, if you are over fifty, it is less likely that 3/4 level TB MSM could have that much effect, but it is possible, given recent news and studies.

I suggest that you increase your dose considerably (1 heaping TB twice a day, at least) and check back. If MSM is the reason that you got this relief, you can probably ditch the drugs altogether within a much shorter period than three months and there is nothing lost.

Also, be certain that you are holding the doses (with water) in your mouth long enough to achieve maximum absorption (no coldness left in your mouth before swallowing). This is key to making this protocol work best.

Further note:

We have no actual studies on this HL* MSM *protocol other than the feedback from two Yahoo Forums: Beck-blood-electrification and Coconut oil. Most of this has occurred in the feedback has come in the past three years,* *(six now with 4th edition)* *and it continues ongoing.

Remember that MSM is "food", not a drug. While we all know that we can eat too much of any food, you know that eating too much food is not the same as taking too much of a drug or even a synthesized supplement that might drive you out of balance. The statement below on peas has been passed on many times in our groups, but it is true.

If I told you that peas could heal your problem and you took 3/4 TB of peas a day, this is where you are. It makes absolutely no logical sense, but these things are impossible to predict. You are likely totally devoid of organic sulfur stores in your body unless you have had access to MSM at some level. By following this, you will have been saved of some terrible pain down the road.

We now know that MSM does affect endocrine/ hormonal balances quite positively, which I did not realize till recently. What I do know is that, had your problem been related to joints, this advice would be almost 100% certain. Since I have never had extended health problems like yours to test personally, I have no first-hand experience with its workings as related to them.

I wish you the best. Please keep us informed. This is quite intriguing. We may be onto something. If it works, "please tell your doctor what you are doing".

I have no dog in this hunt, but my objective is to simply get this information out there and into the mainstream... where it should have been all along... had targeted drugs not become the main focus of modern (allopathic) medicine.. MSM is the most incredible "food" source of Organic Sulfur that I know of. Fortunately, many are picking up on them after many years in suppression.

I am forwarding this to my M.D. friend who happens to be an expert in this area. If you follow the above advice and are totally relieved, I am sure that he will become extremely interested... and I will not be surprised.

Kind Regards, Huuman

A Horse Woman Asks:
>Hi Jim,

I have been (not too closely) following your thread on MSM and am not sure if my question has already been addressed. Nonetheless, here it is...

I have been giving MSM to my horses for years. Vet instructions are to work up to one cup/day and mix in food. If your suggested dosage and ingestion technique is correct, how could this (vet's way) be helping my animals? Since this is a long-standing, (proven?), recommended remedy for joint pain in horses AND readily available in feed stores, could the vets be wrong? If MSM hadn't been useful (in this way) over the years, it wouldn't still be around.
So... I've believed in its usefulness and have used it accordingly for my horses. You're changing the paradigm and I'm trying to make sense of it. Can you provide more information on how using MSM in this way might be effective for animals, but not for humans . . . . and, vice versa.

I appreciate your helping some of us through this confusion.
Love and Laughter, Anne<

Hi Anne:

Simple answer: Train your horses to swish it around in their mouth's :0)

The fact is that any way will help, but I am giving you the best method for humans to ingest it. Also, consider that horses and cows don't have our digestive system. I am certainly not an expert on their systems, but know that they differ from ours considerably from College Biology I. I am sure that your vet can enlighten you on the details and he/she, in turn, may learn something from your question.

Regards, Huuman

>Hi Jim:

I have been taking msm since I joined the Crock_Lakhovsky group in January and first learned about it. I get it from PC Networks and I just don't think it has done anything for me. I take a heaping T every day. I have learned to grind it up in the blender into powder and then I can get it down faster so it is a little easier to take. Tho it is still hard to do. I would like to increase it, but it is like ingesting peroxide. I can only do that so long. So one scoop a day is all I can manage consistently.

Yesterday, I took it with kombucha and it seemed even easier, so that is how I will continue to take it. But I don't think it is going to ever do anything for me.<

Hi Anne:

If one tablespoonful is not doing the trick, double the amount till you get better. The key issue is to get enough... Young people need less to get results. As we age, we simply absorb less... which is generally the problem in the first place. We are not getting enough organic sulfur to keep our joints in order, so we must supplement. Simple... & simple solution. What is hard about scooping it down till you get enough?

Kind Regards, Huuman

Hi Lyn:

No one really knew much about MSM when I first heard about mega dosing. I started out taking much more than 1tbsp/day MSM, but I had taken ½ gm/day previously with no effect.

Is 1tbsp too much? If someone has a mouthful of mercury, it certainly could cause an overload, due to the fact that it frees heavy metals (which is a good thing long term). My instructions were scant. I have given much more thorough/ careful directions here in the past here on many occasions. But OK: ½ gram to start (if your joint pain tolerance is high and you are patient).

However, if you want to get'ter done. Take a tablespoon full and test the water. The heavy metal release could make you feel badly, but it may be just what the doctor ordered if you have no build-up. So, if that has no ill effect, move right on up toward ½ cup/day.

The fact stands that MSM is absolutely one of the least poisonous chemicals on earth... less so than water and it has the potential of relieving a whole lot of joint pain fast... and permanently... Not to mention its wonderful side effects while it is doing its work

Cheers, Huuman

>Lyn K wrote:

Jim, are you referring to 1 Tbsp. MSM or 1 Tbsp. whey? If it is the former, I would definitely not start out that high. This is advice I was given on a different group:

"We always recommend starting at a low level of between 500 and 1000 > mg (eighth to a quarter teaspoon) per day. Below is my ebay source for MSM. $19.95 for 2.2 lbs. plus S&H (and it is for humans). The thing that I look for is texture... too coarse and it is tough to deal with. This stuff is just right IMO.

Age V. Amount of MSM

Lynn<

Additionally, if you have severe joint problems, the recommendation on your referenced site is simply inadequate in my experience. Here is mine:

If you are over 60 years old, 2 oz /day in a 7-day period will help, but I have found that twice that is required... and for as long as it takes, not just 7 days, before all symptoms will eventually stop. In this case, it really is all about quantity. People under 40 absorb MSM better and simply do not need much to effect a cure. I have found that a couple of grams/day often do the trick for most people this age with minor joint pain.

My son is 25 and one gram/day is plenty for him. He has found that missing three days spells pain, though. This is proof that our diets, today, simply do not support joint health adequately. We ALL must supplement.

I have also discovered that MSM for animals never seems to have a texture problem. Human MSM can be too coarse... Strange, but true.

Regards, Huuman

>Hi Jim

Jim, did you mention the brand of MSM you used to get these results? If so I think I missed it. And how much did you say you took before achieving these results - was it 1/4 cup (4 TBS) per day?

Thanks,
Dee<
Hi Dee:

I have bought it on ebay... & gave the address previously. But in those days I bought it from some guy who sold it for horses ***(Ebay & Amazon are both good choices)***

4oz/ day at the max rate. Now I take much less.

MSM never tasted badly to me.

My friend hated it at first, but after she discovered the benefits, she woofed it down like candy and never complained. She says she likes the taste just fine.
Bitter? Maybe some, but it's the cold, clean taste of healing that really gets it... and never having joint pain again makes it taste ever so wonderful. I love the stuff!

Its kinda like when I took cod liver oil as a kid... and got kinda addicted to it. Even now, years later, I still like the taste of it. Too bad we over fished the cod and virtually killed off that industry in New England. It is unlikely that we will chop down all of the trees and exhaust our MSM source.

Huuman

8/12/12
>Hi Jim
I started with 2 tbl spns at a time twice a day and I got nervous like I was having panic attacks so I backed off and am slowly adding more. I'm up to 4 tspns 3 x daily.

I guess some of us are more sensitive to it. I sure hope it's killing off my cancer. I need all the help I can get. Thanks for all the input on MSM. I'm going to keep adding more and more .I'm still having a lot of pain but I know the MSM is helping a great deal. I was doing pretty good for a while but the last couple of weeks have been a little rough.

Steve<

Hi Steve:

Cancer is much easier to avoid than to get rid of, but there have been some good results reported from using MSM, especially in the past two years. It may work just because it frees your body of heavy metals and toxic chemicals. if that is the reason and you have it for other reasons such as a lack of selenium, its not going to do much.

If you can avoid radiation treatment and drugs, that should help though. Also, read Duncan Crowe's site. He has some great tips there. In addition, don't forget how you got here... Bob Beck knew some things.

In any case,
I Wish You the Best,
Huuman

A Bad Solution from a Good Guy
>Jim

My solution to the bad taste is using MSM tablets, which I chew before swallowing along with a mouthful of water with ascorbic acid. The tartness of vitamin C helps with the bitterness of the MSM.

Alobar<

>I am really new to MSM, but I am really new to everything that gets talked about in this group. So here I go with another question: I am puzzled why and how this is an animal feed supplement. MSM tastes so vile and how in the world would anyone get an animal to eat it? I dowse mine with peppermint oil to mask the bitterness to get it down. But if any flakes stay in my mouth, Ugh. How does someone get this into an animal?<

Hi Alobar (and other person):

The key word here is TASTE. MSM only tastes badly to some. I never minded it, but my wife did. Interestingly, she has since gotten use to it and she is not particularly adaptive. Think about what it is doing for you. I actually enjoy the cold part and am neutral to the slight bitter taste.

My golden Retriever got a terrible case of arthritis at age 8 and could not go up stairs. I wrapped it in raw hamburger and he woofed it down like candy. In two weeks he was running up and down stairs like a puppy. His hips finally failed at age 15 after I failed to exercise him in bad weather.

But this is the second part of this supplement... exercise. You have to get it worked into the joints for it to be effective and exercise is the key to doing that.

Tablets? You do not apparently understand the protocol. If you currently have problems, you would have to take about 30 one gram capsules a day to do any good. One the plus side, I have found that if you take enough MSM, you will actually learn to like it. My friend Deb and my wife both learned to love the taste.

Even if I am wrong, the question is: which is worse, sciatica or a bad taste in the mouth? How about a back operation that goes terribly wrong and knee replacement or hip replacement surgery where the bone breaks off, and debilitating arthritis. These are the possible tradeoffs.

Regards, Huuman

Hi Tanstafl:

The key to why is in the chemistry. It numbs the taste buds almost like what a shot of Gin does.

Regards,
Huuman

>Hi Jim:

Very interesting.
I tried it...

It took a while for the cool taste to go away, and it took a lot of switching and gentle chewing to grind the flakes down to the point it was almost all liquid, but I'm happy to say, the taste and after taste wasn't nearly as bad as it is when I add it to water, mix it for a few minutes, then drink it down... why I don't know. <<< ***(this is a common reaction in***

my experience and it apparently is a reaction from the Methyl component of MSM. Tans was the first to report this though... after myself).

Anyway, thanks Jim for the prodding... this will save me a lot of time making the capsules... but I'll still probably need to make them for my mom and dad, who are not as adventurous as I am... ;)

Tanstafl:<

>Hi Jim

I also say that this stuff tastes terrible - and the aftertaste just lingers for hours... so I'm making my own capsules... much better...<

I am interested in MSM. I also like to find the best deal for my (hard earned) money. I like these prices... I would like to understand what did you read that lead you to think that once powdered MSM is "vastly" inferior to powdered form. I have become very leery of marketing claims ... These feed on our insecurities to pitch their wares. Making powder from flakes seems to be a very simple mechanical process. Are MSM properties destroyed by heat?

I thank you in advance,
Frantz<

Hi Frantz:

I have not tried the flakes recently, but I did not like the texture. The cheaper powder works is easier to take in my experience, but both are fine.

MSM is an oxidized form of DMSO. There is no further reduction that I am aware of. Your deduction appears to be accurate and, certainly, MSM is not destroyed by making it in powdered form.

Why heat it? It makes no sense.

Cheers, Huuman

>On the subject of MSM, I have read that Organic Sulfur Crystals are more potent than MSM. Anyone have any experience with this?<

Strange statement that organic sulfur crystals are more potent than MSM is, in fact, a long chain Organic Sulfur that is easily absorbed due to its methane component... Hence the cold taste. All indications are that Goldstein's Crystals are, in fact, a textural form of MSM that, all derived from pine trees. But I am not an organic chemist and my question is, what is the point?

However, going on the assumption that Goldstein's MSM really is 85% more absorbable than powder, is it all that important? Experience tells us that 4 oz a day will heal just about anything when it comes to joint problems with enough duration.

I tried a flaky variety once. The texture did not work for me, so I put it on the back shelf and ordered the right stuff. Within a few months, 4 oz of powdered form turned it around for me. I never expect to need that level of protocol ever again since I listen to what my joints tell me and supplement accordingly. The bottom line is: If Goldstein's is more absorbable, but this protocol is acceptable, then it comes down to cost.

So Goldstein's is not new news, but what is interesting is your reference is the long list of ancillary benefits that MSM yields... good stuff huh? Anyone here ever heard of a drug that has these wonderful side effects? Not in your life. The blood circulation benefit, by the way, is a result of the fact that MSM makes all connective tissue more pliable. So it makes your nails, skin, and hair more pliable also.. perfect sense!

I have been using this *HL* MSM since I was 56 years old after reading an MSM salesman's website on the subject and calling him. He sold it by the rail car load for horses. I have never looked back. He told me about the protocol and the fact that no one in his family ever got sick (!)

The man shoveled me a few pounds of MSM. Since then, has repaired every joint problem that I have ever known including a knee that I tore up skiing at Mt. Killington years ago, the juvenile arthritis that I was born with in my hip and fingers, the lower back problems that I inherited, and a bad case of tennis elbow, and the neck pain that I got from working over a drawing board.

(Note that Glucosamine and Chondroitin (along with egg shell membrane and a host of other supplements) are all Organic Sulfates. Many of these in the Chondroitin family are large and poorly absorbed molecules according to organic chemists. All of them, in my experience, will upset your stomach if you take enough to heal you. They are all much more expensive to produce and sell, if that is an issue.)

LOL on Small Amounts of MSM
>Hi Jim:

I am very, very new to supplementation . I had preferred "Whole Food" for a long time until I finally understood how unpractical it was to find "whole foods" and the systematic depletion of our soil (in the USA at least) thus that our foods no longer are reliable a source of certain components/ nutrients , e.g. Brazil Nuts analysis reveal the paucity of selenium from the Brazil Nuts available on the market ... but I digress ..

I am not sure it is arthritis but my knees have become quite painful when I run to such an extent I have almost given up running and use elliptical machines for my daily workouts... My knee's creeks quite audibly when I crouch for example ... So repairing my joints is in order before other parts start creaking too ....

What is the minimum dosage for MSM for a person to notice a difference? I just bought some MSM Tablets and noticed that they are 1000 mg per capsules I would need to gobble 112 capsules everyday to equal what you take daily (4 oz) .. An Ounce is 28 grams or 28,000 mg .. Am I missing something here?
Frantz<

Hi Frantz:

Whole foods are best when you are getting enough nutrients as you suggest, but if you are having joint issues, your body is telling you that **you are simply not.** In fact, many other diseases may be telling you that including heart disease and cancer... listen to the Dr. Seneff video for that story.

If you were twenty-five years old, you might get results with a couple of grams/ day. However, most people with bad joint problems are older (my son and I were not). So if you are 50 or older, that amount may not make a dent. In any case, start off slowly. MSM will release heavy metals that are bound up in your body fat. They do little harm there, but when they are released quickly, they can make you very ill. Yes, 112 capsules a day would be OK, but capsules are way too expensive and are not well absorbed.

Since we are all different, there is no set amount. Even a gram/ day is better than nothing (and you are going to spend a lot of money in the process). I buy mine in 5-10 lb. plastic containers on ebay or Amazon for about $40. It is specifically for marked for horses.

Cheers, Huuman

>Agnes<

Jim, My daughter has rather severe problems with her back and hip joints for which she has to go to the chiropractor constantly for adjustments as well as she has tried acupuncture. Now looking at various types of exercise that might help. Do you think MSM will help? Please advise where you purchase your msm, how much and how you take it. Comments from anyone else will also be appreciated.
Agnes<

Hi Agnes:

I certainly exercise. Add Duncan's whey protocol and make sure that she gets plenty of exercise. You also might add to that Interval training (see Dr. Mercola's site for videos on that). Without exercise, the nutrients can't move into the joints (as I have warned previously)..

The point is that Duncan's whey alone may correct a mild problem, but this *(HL)* MSM protocol absolutely works in most cases if you stay with it. I have tested it on myself and everyone that I know with joint problems who is not hung up on allotropic medicine. It has **never failed** as far as I know. As previously stated, there is no such thing as an organic sulfur overload. Too much is cheap insurance, and you simply pass whatever you don't need.

Finally, even with daily MSM, MicroLactin and Whey supplementation, I still hyper extend a knee, elbow, etc. occasionally a bit playing tennis and the answer are always to take a couple of heaping tablespoons of MSM as soon as I get home. From there, I just move on and forget about it. So far, in ten years, recovery is always immediate. Life is good.

Read my past posts on amounts, etc. There is a lot to this.

Regards, Huuman

>Just to add to the general information, I have been taking MSM for over a decade.
I've found that results coincide with dosage. I've used up to two tablespoons at once.
Usually a teaspoonful is enough.
Chuck<

Hi Chuck:

You have not told me anything about yourself. There are four important aspects to consider that effect your joints: Age, Diet, Weight, and Level of Health<

If... you are 25 yrs old, in excellent health, not overweight, and you happen to enjoy lots of homemade soup made from scratch: a teaspoonful is enough per day is probably all you need right now.

However, I suggest that you not keep this as a rule forever, because, chances are, you will likely age. Unless, of course, you have been reading Duncan Crow's posts for a number of years.

Regards, Huuman

>Jim remind me again-- what is the web site you use to buy MSM?

Thanks
Ron<

An Important Find From bG (and a total surprise to me, really)
>Hi Jim::

Interesting that MSM has even more benefits than I ever conceived:

As in cancer also, the byproduct of fermentation acetaldehyde, blocks the function of the hormone like prostaglandin E2, the anti-inflammatory signaling hormone. This is why using msm to neutralize it followed by a prostaglandin simulator such as **cod liver oil** along with a saturated fat which improves this process is important to defeating candida and cancer for that matter. I know all this because **my daughter's tumors are shrinking and dying, have changed colors since starting msm and I wanted to know why.** Took me several days of research, but think I figured it out.

bG<

Hi bG:

Always a great joint cure, this is one on many recent posts relative to MSM as an antifungal candida and cancer preventive and cure. I am glad that there is so more interest being paid to MSM, in general, today. There is no question in my mind that it totally changed the course of my life when the guy on line, who sold it by the rail car load for horses, offered me a few pounds for my many joint ailments and informed me of this *HL* MSM protocol.

I am happy to report that after a couple of months on it, I have not had a single joint problem. One of the other great benefits is that it makes your circulatory system more pliable... imagine that! No likelihood of ever getting heart disease.

It is getting to the point that I am expecting MSM to cure everything... then, where it is unsuccessful (as a topical cure) DMSO, its sister, will handle the rest. This is an amazing pair indeed.

An additional benefit that I recently learned is that MSM is one of the least poisonous chemicals on earth and even less so than water.
I think that a big clue that has been overlooked in the cancer conversation because it is "so simple" is our need for organic sulfur in our diets.

Regards, Huuman

The following questions, answers and sometimes expert commentary mostly followed this actual written book. They also immediately followed the first Dr. Stephanie Seneff interview with Dr. Mercola which profoundly influenced the author's thinking and added greatly to the Sulfur part of the MSM Story
(http://people.csail.mit.edu/seneff/***).***

It was always clear that HL MSM was magical, but her enlightened information lifted it to an new level pouf understanding as to why it worked so well..

54. We Are Learning More - More recent Discussions

>**Steve** ***(post from 12/12 before I had fully researched for my book)***
Since you sent me your E-Book around 9 months ago I started off slow and you told your pea story so I started upping the dose gradually. I went up to 2 tbl spns 3 x daily.

- **My hair got thicker and healthier.**
- **My finger nails got healthier and needed clipping every week.**

- **I had a balding spot on the back of my head, One day after taking MSM for several months, I asked my daughter if it was still there. She answered "Were was it".**
- **An ND friend told me that hair is linked to the liver and it meant that my liver was functioning better.**

I wouldn't want to wake up in the morning without it.
I just ordered 5 lbs. It has arrived, but there were several weeks that I had to back off for fear of running out completely... and the benefits slowed considerably.

So glad I can go back to my normal dose of about ½ a cup per day.

Many Thanks,
Steve<

Hi Steve:

Thanks for the feedback... ain't it great? Bald spot disappeared... that is a new one. Maybe you have reported the first of this ever occurring. You could have made history. Since I never had any bald spots, it was never something that I could have proven on my self. The same is true
with heart disease and cancer. There is every reason to believe that MSM will cure these also, but neither me nor any of my family has had them to test since discovering this protocol.

By the way, once all of your possible health problems have disappeared, you should be able to slip back to twice a day. However, doing ½ cup/day or more has no real downside and it sure is cheap to do.

Given reports like yours, I really can't see how anyone could ever avoid using MSM and following my protocol. Its just too good to ever pass up.

Of course, your friend is correct regarding the liver, but the liver does just about everything, at some point. Rightly, we worry about our hearts, because of the dietary problems that modern medicine has passed along, but this is really the organ that we must pay attention to first. As you say, MSM nourishes it just as it does joints and arteries. It restores just about every bodily function while keeping us clear of toxins. The implications are huge in this society riddled with chemical toxins like mercury... and the resulting cancer that they cause.

By the time that I get around to actually publishing this, I expect that there will be no bodily function or disease that MSM has not shown to improve in some way.
Dr. Stephanie Seneff basically told us that in the Mercola Video, right? Stephanie is just one damned smart cookie, but she really needs to read about my protocol also. Knowing that being low in Organic Sulfur is huge and knowing how to take care of that problem easily are two really different things.

Stephanie obviously does not have a clue when it comes to MSM and neither does Dr. Mercola. But how could she unless she comes here or reads my posts in the Coconut Oil Group that I frequent? Most assume that this is common knowledge just because MSM has been around for so very long. It is not... and if it were, it would be in every health discussion on TV.

When you combine MSM with a Carb and sugar free diet, and Duncan Crowe's Whey Protein isolate, it is possible that we have something that will stretch an average lifespan to numbers that no one has previously considered even possible in modern history. We could be back to Old Testament lifespans like 800 years, who knows? Becoming 120 years old should be a piece of cake... bad comparison... but you get the picture. If you

have not read Duncan's anti-aging site you must.

No doubt about it, when stories like yours finally get out, it would literally change the face of modern health. I have never encountered any single change, anywhere, that yields such results. ***But just getting rid of bald spots is, in itself, big enough guys, right!***

Below is a post that I wrote for the Yahoo Coconut oil group that I believe, along with the referenced Mercola Video, begins to describe the optimum health story pretty clearly:

The bottom line is that I increasingly believe that you can set yourself up to be an almost zero cancer risk (barring an extreme problem like a near radioactive toxin source) as well as heart attack risk as Dr. Seneff suggests and possibly a zero risk for all of the other terrible diseases that Dr. Seneff touches on in the Dr. Mercola interview... plus the ones that I suggest in my comments.

To this, add avoiding colds and flu if you supplement with D3 in winter and Astaxanthin and even avoiding sunburn (!) when you overdo D3 in summer (this has the added advantage of allowing us light-skinned folks to live in the Virgin Islands full time without turning into prunes). And this can be accomplished in just a few weeks by going on *HL* MSM Protocol, then converting to Duncan Crow's Undenatured Whey Protein/ low Carb/ no refined/ low sugar Diet.

It is just too simple to be true, but the last two years have led me to increasingly believe this as the facts are revealed. Combine this with the weight optimization and anti-aging that this diet suggests with all of their benefits (well maybe...) and things get to be really interesting. This will not likely take long to prove out here on the web now that this suggestion has become clear to us.

I have noticed that things are just getting easier and more simple as the world situation, according to the mainstream press, approaches a collapse. I am excited. Can it be this easy? Probably not, but it is compelling and worth the effort to give it a try.

By the way, cigarettes did not apparently kill the American Indians and that suggests that they would not kill you if you got them unprocessed and natural as they did. Our tobacco companies have apparently made them much more dangerous and more habit forming through such devices as chemical growing methods (that poison you), "a silly millimeter longer, menthol, and filters". We may never know the answer to this one, but then, I am not recommending using tobacco and have never had the habit.

Most likely, an example of a similar industry is the one emerging in pot smoking. I predict that they will end up making pot very dangerous as they commercialize it, if it becomes a regulated industry like the cigarette companies. They want to increase their market and will discover how, even if it kills you. The bottom line is: let the Government and Commercial Industry together and into your life... and things must eventually go this way. This is why our founding fathers (TJ especially) had such distrust for large government, regulation, and taxes... and why Government loves big corporations. They manage to mutually benefit from their growth.

Cheers, Huuman

>Hi Jim

Got a spec of good news: Samoans smoke like fiends, Hawaiian don't. Yet, the lung cancer rate in Samoa is only 1/10th the rate in Hawaii. Dr. Joel Fuhrman is my current hero, ok so laugh out loud, he said this on one of his videos, that the Samoans eat lots of fruits and vegetables, but the Hawaiians don't. There may be other factors hidden in the statistics, of course. But I've changed my diet a lot since listening to him, not sure if will help with cancer should I get any since I smoked for 20 years about 25 years ago I quit...but I have lost almost 18 lbs. in a few weeks while being stuffed!!! :) BG<

Capsules Work, But More is Better w/ MSM
>Hi All:

You can take MSM capsules and get enough, it has changed the quality of my husband's life who has **stage IV brain cancer.** He takes pure encapsulations MSM 850 mg 3 times per day. All this talk about not taking capsules is misleading.

Carol<

Hi Carol:

Your assumption is hardly the truth. A little knowledge is a dangerous thing, in my opinion and *especially in this case.*

When I first discovered this *HL* MSM protocol that I have recommended here for sometime, most of my body was hurting. I was taking 1000 mg in capsules twice a day. That amount gave me some benefit, no doubt, but I was getting more sick, *daily*. Certainly, 550 mg more, as you recommend, would have added little to my deficiency.

I have been actively getting this out for about three years now, but have told plenty of people about taking *HL* MSM from the minute that I got better... and that healing occurred in about two months. Since that time, many people have listened and many simply don't read things thoroughly or have not heard all that I am saying. In addition to Chuck's earlier comment, here is an example:

Brian, tennis friend, age 50 (then), partially heard me in 2000. He, a nurse, had a really bad knee problem. In fact, his problem was so bad that his Dr. told him that knee replacement was his sole option. Brian took two 1000mg tablets/day and was completely cured within a month. Obviously, Brian's deficiency in MSM was not that great. He is grateful and never fails to mention how he got better when he sees me. I am happy for Brian, but he was lucky. Since he has gotten better I have corrected him, because his problem will recur unless he understands the *HL* MSM protocol completely.

Your husband was lucky to have benefited, but his cancer may has gone away completely with enough of it. While an organic sulfur deficiency may have only been a contributing problem, we will never know *if it was "the cause" as we now know it could have been.* Even a small deficiency in organic sulfur, as he proved, can have terrible consequences.

As Dr. Seneff related in her interview, Organic Sulfur Deficiencies can cause just about every disease imaginable, including Cardiovascular, Cancer, MS, etc. Everyone should hear her interview at least once. I will always repeat this, because her interviews are key to knowing what *HL* MSM does.

Seneff is a marvelous observer, but she does not even know about this *HL* MSM protocol and due to the prejudices of her (scientific) group, she may never learn about this. Taking *HL* MSM is efficient and cheap and it wards off or cures people of all kinds of problems, but as Dr. Seneff says, "Joint diseases results because your vital organs need the organic sulfur and steal it. If it did not, you would die of organ failure (heart failure?)". This is a profound fact that I was not aware of till this interview. Looking back, I am glad that I suffered so many years from joint ailments rather than brain cancer or a heart attack.

I believe that most everyone here would agree with my assessment above, but then, I have absolutely no fear of dying. I have been there and loved it. Death is an especially excellent ending for a tired and sick body IMHO. Nevertheless, I have more work to do and some of it may possibly even entail keeping you well, Carol. However, I get will this out and you may do what you wish. Enjoy that freedom. It may be fleeting.

Finally, there are many other ways of healing people and maintaining wellness. Duncan tells you about whey and some of his other products. He has thoroughly researched them. Listen to him carefully and do not minimize what he says as you have attempted to do here, possibly unknowingly, with me. You may mean well, but you are generally incorrect on this. 850 mg three times a day is seldom enough to cure anything.

Kind Regards, Huuman

Think about it Laura:

HL MSM is just an elegant delivery system for organic sulfur. You body rejects too much of it once it has had enough.

However, if it does not have adequate amounts for your vital organs to function properly, **your body steals sulfur from your joints. This, so that you don't die from that organ failure.** The result from this is that your survive, but with joint failure and arthritis. To paraphrase Seneff, Better a bad joint than a dead organism. That may be your body's opinion, but mine was, "I am pretty sure that I'd rather be dead than to go through this sciatica pain again". I think that most would agree if they had my experience, but few ever actually kill themselves over pain.

This is a very simple explanation for a very complex process that has alluded doctors virtually forever, but from all that I have read and discovered, especially recently, it is pretty accurate. Inadequate organic sulfur has been very closely connected to cancer, heart disease, most autoimmune diseases, and, of course, joint failure.

HL MSM is the quickest way to get back on track if you are deficient and Duncan's Whey Protein Isolates certainly will help keep you up to snuff. However, my experience is that, as you age, MSM is the only way to stay on top of the issue even if you are not on the SAD diet.

The bottom line is that it should never affect you badly, no matter how much you take.

Regards, Huuman

All Natural Remedies Reinforce
>Hi Jim:

That really makes sense about the body stealing organic sulfur from other places - i.e. joints...
My herbal training tells me that MSM works better in conjunction with organic rose hips = good source of Vit C. have you ever heard of this?... 'It aids in the production of connective tissue and for pain reduction' - aka decreases inflammation.

Thanks,
Don<

Hi Don:

No, but these remedies all reinforce each other. As Jon Barron says, Vit C is best taken when it is bound up in fruit with all of those other cofactors. I stopped taking unbound C years ago when I first heard that.

Where do you get your rose hips? I used to gather them up, but got lazy and my bushes died.

As far as joint health, MicroLactin (Swanson Vitamins sells it) has also been a great supplement along with MSM

Cheers,
Jim

>Hi Jim:

I believe this is the same video on you tube. There are 7 parts, should we need to watch all of them??

MAJOR thanks to Jim for posting this MSM info. Jim, if possible could you post your complete protocol again? I'm trying to help my Mom overcome her joint pain. I've copied about 10 of your post via the yahoo search but I didn't see the total protocol.

Also do you think the Animal MSM on Amazon is a good one? Keith<

Hi Keith:

My answer is:

It is, and you bet!!! Watch it several times.

I have gone back through my messages from this and other groups and put them into a somewhat compacted version, but it is still too long to post online.

I just sent it to you... and you are welcome, just as I am grateful to the horse med. salesman who told me about it years ago. I would be living in total hell by now ***(if at all)*** if it were not for that man. But then, chronological age is certainly not at all an accurate way to compare people's health or life engaging abilities as Duncan Crows will tell you.

Keep this in mind: the real human life cycle is 12x12=144 years, not the puny 80 or so years that we normally think of as a life span. Science relates it in telomere length, but we are virtually killed off by mainstream medicine just because they do not get it. You and I must teach them that they are wrong.

Yes, the animal MSM (for horses) is high volume. That makes it high quality. That is actually my first choice and is what I am now using *(and have been since day one, mostly).*

Kind Regards, Huuman

>Have another question.. can or does MSM affect thyroid function? Was dx'd with hasimotos about 5 or 6 years ago... has been badly treated as no one really understands the disease, have been having problems since before I started the MSM, but it really went hyper to the point I had to stop the Levoxyl. It is now settling down, but one of the symptoms of hyperthyroid is anxiety, IOW, squirts of adrenal hormones.

I was actually told the thyroid was burnt out, at this point pretty sure they were wrong. And yes, I tried natural thyroid and did very badly on it. I am really at a loss. Going to see and endocrinologist this week, don't expect much, but its necessary given that Hyperthyroid can really cause a lot of damage. I just really don't know what to do or think at this point. It is amazing how well I feel given how many things are wrong.. its just so overwhelming. I really feel I need a definitive dx and then I can look at adding alternative therapies.

Sorry for all the rambling. Just been so frustrated.

Laura<

undenatured whey is not enough
Hi Duncan:

I'll check out the PubMed studies. In the meantime, its good to know that you were ten years ahead of the curve on fucoidans!

I assume that it is because it is related to Whey?

By the way: I have discovered that, at least for me, **undenatured whey protein alone does not supply enough organic sulfur** to keep me out of joint/ arthritis problems. I had avoided daily doses of MSM for a couple of months on the possibility that it would, but last week, playing tennis, I got a foot cramp (first joint problem I have had in years) and last night I got a bad left hip joint pain (felt like the right hip joint that set me up for sciatica some years ago).

Two scoops of MSM put them to bed (as I knew that it would), but it also taught me a lesson that I am passing on for any active 70 yr old here...stay on lots of MSM if you want to remain active. I expect to stay on the courts for another thirty years at least with the help of Whey and MSM plus your anti-aging program. And this is a study that I don't mind participating in for free. It is actually very enjoyable!

Can you imagine a guy 120 yrs old (in the year 2061) who still has a good tennis game, (etc.)? What record book would that make? It sure seems possible.

Duncan... Again.... it is great to have you around my friend!

Cheers, Huuman

>Hi Jim; glad to help. My fucoidans page is 10 years old, so there are probably newer studies of interest in PubMed :)

all good,

Duncan<

Antibiotics & Dairy Cows
General note:

As Duncan has said. and I agree: Dairy cows are always free range and always raised on grass (with some silage), but the more important thing in my mind is: Are they using antibiotics? I have discussed this at length with my favorite local dairy farmer (see Chapter 35) and he tells me that he simply can't afford to give his cattle corn or grains except in very coldest weather when nothing else is available.

Beef Cattle are another story. They are generally fattened up using grain and not allowed to exercise.

--- In Beck-n-stuff@yahoogroups.com.

Hi Duncan:

You input is always invaluable to our health discussions.

Thank you for being here!!!

Regards, Huuman

MSM for Horses is Best
10/15/2011, Laura Cody

> thanks,

You're the first to acknowledge that the horse MSM is probably the same as the human MSM and way cheaper... I found some at Amazon... AniMed brand about the same price as your guy...

http://www.amazon.com/AniMed-Pure-MSM-5lb/dp/B000HHM9WI/ref=sr_1_1?ie=UTF8&qid=1318726519&sr=8-1

Whatchya think?

Laura C.<

Hi Laura:

I think that Mike "the Health Ranger" Adams, is a little over the top on this one in his warning, but he is essentially correct.. I liked Mike's page a lot better when he was small and did not have a staff writing for him. He essentially lost my interest when he grew successful.

MSM, the organic formula for oxidized DMSO, is best purchased from a fast-moving supply houses. As far as I have ever heard of or am aware, it ONLY can be made from oxidizing DMSO (This is wrong. Now it is made from organic chemical processes 11/13/17) and since it costs almost nothing to do that, there is little worry that as long as it does not sit long, it is good stuff. That is why, years ago, I was advised to buy it in bulk from a supplier for animals... specifically horses, where it has been used effectively, and always in bulk. Since animal suppliers actually sell it by the rail car load, that turns out to be the most correct choice, but any fast moving supplier is OK. I have bought it in bulk on ebay and from horse people.

You know that it is good when you get that "cold" feeling in your mouth. There is nothing else like it. You can't even call it a taste, per se.

Pills have additional problems to those mentioned: there is just not enough there to do anyone any real good; if you don't scoop it in your mouth and follow it with water, you don't absorb near enough: the MSM has been handled too much even if it has never touched human hands, it picks up any and every chemical around it which is why MSM is such a great detox chemical: and finally, pills are far, far overpriced. If you buy enough MSM in bulk, you could supply a small city for a few thousand dollars for several months.

Their study only confirms what I was told many years ago by that horse supplier. Go for bulk suppliers and take as much as you have time for. I might add... use it quickly and if you can't, move it into a tight glass container where it can't pick up chemicals.

Some MSM History
From Huuman (snipped from another site):

>"Here is the crux of our Study: Most MSM as commercially available does not contain 34% sulfur. Some contain no sulfur by observation. Like the fertilizers that we believe have broken the sulfur cycle the additives of packaging block or neutralize the bio availability of the sulfur contained in the MSM.

Organic Sulfur is *basically* MSM precipitated from DMSO into a coarse crystal flake and not processed any further nor does it have anti caking or flow agents added. When the crystals we use in our Study are pulverized the effectiveness is reduced according to our Study members. Our Study has had to ship "powder" rather than crystal Organic Sulfur on two separate occasions. Many of our Study members had been taking some form of MSM prior to joining the Study, **the least effective are those in pill or capsule forms.**

When Dr. Stanley Jacob and other researchers began their MSM research it was with bio available sulfur. The resulting marketing of MSM has resulted in a less bio available form of sulfur. The actual manufactures of MSM are few while the number of resellers are very numerous but unfortunately we believe their products are contaminated by the additives of packaging just as our foods have been contaminated by over processing and chemical additives. Sulfur is very interactive and can form sulfates, sulfites and sulfides with most of the other element we find on the Periodic Table."<

Looks just fine to me. As you may know, I started this *HL* MSM for horses, here. The guy who sold it to me told me not to get carried away with initial quality. The problem is that it can pick up any local contaminants around it. So when it is not handled as much (he sold it by the rail car), you are getting the most fresh and least possible contamination.

With that in mind, the best place to store it is in a big glass jar with a top. I keep mine in one and ladle out smaller amounts... enough for maybe a week for the wife and I, then refill. Certain kinds of plastic could really be a bad thing to keep it in for any long period. We'll probably go through 5 lbs. in a couple of months. If you are going to keep it longer, I'd definitely purchase a special glass container for it.

Regards, Huuman

10/30/12
>Jim, Where do you purchase MSM for horses?

Theta<

>Please explain what you mean by the phrase msm. Does this mean re taking it with other supps. Or mixing it in a cream form with other items.

Thank you,
Lymeover<
>Does anyone know if there are other alternatives for a stress test? I've found a couple of sites mentioning cardiac MRI but the doctor's office says this isn't enough.

Thanks so much.<

10/30/12
Hi Lorelei:

As to buying MSM, Ebay is a good start, but most any place that handles animal feed and products sells it. I just found a new one on line that sells 5 pound lots for $35 with S&H.

Five pounds when taken properly should not last you over 60 days if you have any joint issues. Any joint issues are a great reminder that you are sulfur deficient. Think of them as you would the red light on your dashboard. If you don't take care of that red light, a heart attack could be the next issue.

Be careful what you store MSM in. It will absorb anything near it. Glass containers are always the best answer if you keep large quantities very long. The good thing is that it will absorb any harmful chemicals in your body.... especially things like mercury and heavy metals... nice to know huh? Does it absorb other organic nutrients? I can't say, but given what we are hearing about its healing qualities, this would not appear to be the case, but there is much to learn.

Topically, I would suggest its parent, DMSO as a curative. It is also pretty cheap. But be careful, DMSO absorbs things even quicker than MSM and can be very tricky to handle. MSM is the oxidized form of DMSO.

Also, don't forget your whey protein isolates. They also contain significant amounts of organic sulfur (and read Duncan's site on this... it is awesome). I currently eat/ drink about 5 lbs. of whey and 2 ½ lbs. of MSM per month. At one point, I was consuming over ½ cup of MSM per day... that is a lot of MSM, but you simply can't get too much of this nutrient. I am now 70 and I have "zero" health issues... really, for the first time in my life.

Lorelei:

People now say 70 is elderly, but age is really not the issue for the most part. Its nutrition and exercise. As we age we just need more and better and most get less and worse. The real lifespan is somewhere around 130 and that is currently thought to be defined by telomere length, but whatever it is, there appears to be a defined limit far beyond what we currently think of as old age.
I have a good friend right now who went though the operation that you are considering. She is wheelchair bound and will likely never recover (three years now) ***(she has since passed).*** Please put your Mom on MSM and Whey before you allow this dangerous cement injection operation (which will get in the way of the natural healing process). Watch the below mentioned videos and read Duncan's site carefully. If she is diligent, you should have her totally well within three months and there really is no downside. There are no guarantees, but this is what I have found so far. Remember these nutrients all work together and there are plenty more that can add to the healing process, but consider this your core. So add such things as liquid colloidal minerals (NOW is good)... and especially selenium, magnesium, and vitamin D3. I also love a little Megahydrate with every drink and MicroLactin (Swanson vitamins has it), if she is not lactose intolerant, as an adjunct to MSM but there are plenty more. Oh, and forget the stress test IMHO. Her heart and arteries will improve with her back and all other joints. Its all closely related. Also, exercise is always key to getting these nutrients into the joints... any that she can do is good and with as much motion as possible. When she gets better, have her take tennis lessons.

If you have not watched all seven of the Stephanie Seneff/ Dr. Mercola "You Tube" interviews where organic sulfur is discussed in detail, you must, admit. The woman is brilliant.

Huuman

Borax

>There are a lot of ppl posting about using 5 mule team borax at earthclinic. Com<

Yes, as Duncan recently pointed out: Borax is often used as it is considered less poisonous. Go back and read the "Borax Conspiracy" website. I am certainly no expert, but I am all over learning about this.

Borax is about half as bio-available, while boric acid is reported not poisonous even though the label will say otherwise... the reasons that they state are the "Conspiracy". As I stated earlier, it all of that worries you, just buy the Swanson Boron supplements and be done with it... you are looking for 3mg of working boron by whatever source best suites you.

Thus far, I have taken about 3/8 tsp/day of boric acid and have suffered no ill effects. I don't expect any either when I get to the 1 level tsp/ day max recommendation (therapeutic dose), but I'll let you know. Since I have no known health issues, I should be a pretty good test specimen on this one... but I am also in no rush to get to that level.

Since boric acid kills candida (thrush) by all accounts and it makes sense, I like this choice. It is also good topically for athletes foot and other fungal infections and it kills wood rot which is a fungus. Its all common sense and it all reportedly works as claimed. If you have ever had thrush, you know how nasty that one is and I have. Going further, if there is any truth in the claim that cancer is a fungus, then it may have anti-cancer effects. The point is, why not?

Regards, Huuman

>Hi Jim:

> Why no Juice? I rarely drink juice *(alone)*, so I am curious. It is often recommended on MSM sites.
Laura<

Hi Laura:

Well, you can cut back and go up slower, or stick with the 1 teaspn. and see if it goes away.

Water is always helpful for detoxing, but the protocol that really works is: **scoop it in your mouth, follow with water, swish it around for as long as you have patience for, then swallow.** No juice. Never swallow it with water or any liquid as a chaser. Mike Adams gave you bad advice in his article. There is wisdom in this protocol *(as it was given to me)* and it works best. You absorb it in your mouth, not in your stomach. It is simple.

Hardly anyone understands this protocol that I was taught. Juices are not the problem. However, taking MSM any other way can be, because it means washing it down, rather than absorbing it in your mouth. Few health gurus get this distinction (including Mike Adam's writers). Those who have been successful doing it any other way are either young, where any amount will do the trick or they are taking even more than they would need if they did it this way. Can you imagine anyone taking more than ½ cup/ day? I can't. I simply don't have enough time in my day to do that.

Regards, Huuman

12 22 12
Why MSM did not Work
Hi Steve:

Just to let you know, that man was simply not taking enough MSM to do the trick. People all over the world have been taking MSM for the last twenty years, and for the most part, reports have been negative unless they happened to be twenty year olds.

I have said this before here, but think of it like this: MSM is food. If I told you that peas can cure cancer, would you eat one pea per day? That is about what they are doing and why it does not work for them. All organic sulfur sources would cure joint problems if you could take enough of them. But they generally cause indigestion. MSM is very unique in that it is tolerated by most in huge amounts and it is virtually never poisonous. As a chemist noted a few years ago it is less poisonous than water, but there are exceptions to every rule. A few can't tolerate much of it. To them I suggest many small "doses".... but we are not really discussing doses here, we are talking servings. Its food!

I have been well aware of Jim's work since he saved his guide's life with it... what twenty years ago now? But there is no earthly way, no matter how good it is that MMS should have any nutritional affect on people's joints. Also, like Duncan, I am wary of taking any strong chemical day in and day out even in four drop amounts. I have a bottle of it in my refrigerator, but its probably no good any longer. It cured my brother of a contagion once and I know that it works.

When you think about it, MMS is being used as a drug no matter how you cut it. It just happens to be an unconventional one. Our general bent here is toward nutritional healing and that, we know, makes sense. MSM can be combined with other joint supplements and it just gets better. It can be taken in huge quantities and it will never poison you. That is how wellness works. You get more than you need and you just pass it. Not to be negative, but try that will MMS and you will die and Jim Humble knows it. The stuff is poisonous in fairly small amounts, but it will kill pathogens if you don't take too much at a time just as you suggest. But many of the drugs administered today are deadly when overdone or even mixed with others. In fact, most people that read this here know that many legal drugs are deadly when taken as directed... which is the rub. It's a case of collateral damage if you are the one of the unlucky ones.

Taking your Motrin in place of Tylenol is just the opposite of where I place my interests. All non-steroidal inflammatory drugs, including aspirin, are quite harmful over time and all will all attack the most important organ in your body, the liver. No drugs, period, for me thanks. I'll stick with wellness and food.

Jim Humble's MMS2 is still another story. I don't believe that I'd ever touch that stuff in any amount, ever, nor would I ever recommend it to anyone.

Duncan's wife is doing well on MSM because she is still young enough and because she is getting a good additional supply of organic sulfur with her whey. Chances are really good that she will have to step it up as she ages as I have learned is required if she wants to continue to avoid pain (which is our natural early warning system)... even if her aging is slowed with Duncan's anti-aging techniques. It happens to the best of us. My wife is 58 and she stays comfortable with a tablespoon twice per day now. She knows how the system works now and is careful to stay ahead of the curve. There is never any reason to need relief from joint pain... just up the serving.

By the way, I do take one drug... Selegiline HCL at a 1/4 tab per day. Thanks for that one from Duncan and his reference to the Canadian Pharmacy. It apparently increases my balancing abilities. I just read where the average 60-69 year old could only stand on one foot with their eyes closed for 3 seconds before they fall over. I can do that for 17 seconds before I put one foot down (and I never fall). That puts me in the 20 yr old range.

Regards, Huuman

>Hi Jim:
>To add to the controversy, I have recommended MMS to a man who took MSM
>mixed with vitamin C for years and who had developed a hip problem. After
>taking the MMS, the hip problem went away. I suspect that bacteria in the
>mouth, either from gum disease, root canal, or old extraction sites
>(cavitation) is responsible for that (arthritis). Although MMS may help to
>keep the bacterial infection at bay, it does not remove the 'root problem',
>which may be due to bad dentistry.
Steve<

Some History on the AHP
1/9/13
Hi Steve:

I certainly believe that you can... and will get past this, if you continue to do your part and it
appears that you are so far. Taking all these elements together (High Level MSM, or HL MSM + Undenatured Whey Protein Protocol, or UWPP + Low Carb Diet + Energy (or raising your spiritual energy levels) that I am now calling the single term, Anti-Aging & Health Program, or AHP in my book), you are IMHO getting the most powerful level of the healing art on earth (and beyond).

Duncan has been polishing his Undenatured Whey Protein Protocol and the story behind it with Glutathione, ever since I first met him online about 15 years ago, but I am really just beginning to understand what he has been telling us with regard to its contribution to liver function in just the last few weeks. This, as I have gotten back into expanding and editing my multimedia ebook (because of your interest and the others here).

Liver health is huge when it comes to beating cancer back into its orderly place (many here have said that we all have cancer, but we keep it in line. Even though I have repeated that saying on occasion, it makes even more sense to me as this story unfolds). I really started working on getting a handle on cancer when my two lovely aunts and my cousin died of breast cancer within about a two-year period about six years ago. They were all three beautiful women and they were, I am discovering, not more than half way through a healthy lifetime, at most.

My cousin was only in her early 30's, so she had a potential of over 5x her years. We can accomplish a great deal more spiritually with longer lifetimes. Currently, we just barely get a glimpse of how this life really fits into the overall picture and then we die... So dying early is just not efficient!

You and everyone else are welcome to the latest update of my ebook... just ask. It only takes me a minute to respond and send you a PDF of it. As I search down links and read what the "big boys" like Dr. Stephanie Seneff and Dr. Chris Masterjohn are telling us about sulfur and lipids and then feeding that information back into the liver story, I am amazed at the beauty in all of this... it is way more than just a science. It truly is a healing art. We are painting a scene that expresses total wellness in an new, but also very old, and lovely way.

There are several new suggestions in my book that I suggest you add to your diet program, but are a few: One, you must get plenty of Vit D3 (and you don't mention that in at least this post). Also, Moringa is showing me that it is far more than just hype. The Moringa Leaf is a natural stimulant that never lets you down and it is high in D3 and astaxanthin, but just about everything else you can think of and probably things that no one even knows about that help improve health. Duncan will tell you about the anti-cancer qualities of Selenium and with good reason... don't leave it out my friend.

Basically, you have to feed your body (and liver) the highest and best food on the planet... take no chances. Then, you have to be sure to maintain your digestive ecology using things like the Prebiotic Inulin that Duncan has always spoken of and eat real cultured food like Natto to keep things working well.

Regards, Huuman

>Hi Jim:

Oh BTW, I have no problems with the taste. I took MMS the old way for a long time. If I can drink that, then I can drink anything. I only taste sulfur slight metallic taste...not so bad while swishing. Tastes a bit nastier when swallowing which surprised me since your taste buds are in your tongue. With TMJ the swishing motion causes pain is all. I had tried oil pulling with coconut oil but again the swishing did me in but that was for 15 minutes at a time.

I have tried and done many things also on this journey. I use a chi machine and a infrared sauna though not as often as I should. We moved out on my sister's 40 acres 3 years back and the work related to growing our own meats(rabbit, turkey, chicken, quail and am started to grow fish in an aquaponics system)/eggs and veggies along with wild foraging (we have 1000s of blueberry and service berry trees and blackberries grow everywhere when we get rain has also taken it's toll on my body. I gather cactus, spiderwort, toothed savory, acorns for acorn meal, Hawthorne (make my own Hawthorne tinctures), beautyberrys for jelly. We eat everything from scratch. We have just gotten 25lbs of quinoa seeds to replace rice and potatoes except for sweet potato. We do eat wheat products since Ed simply can not go without bread and his sweets but I do make my products from freshly ground einkorn wheat. But at least it is made with whole grains and the ice cream is from raw milk. I could get by with coconut peppermint patties using stevia or "just like sugar". I use chia for omega 3 along with LSA (flax/sunflower/almond) meal that is shaken on meals for good fats. I generally only use coconut oil or olive oil.

Ed had prostate cancer and we went to almost all raw (hallelujah diet and drank carrot juice by the 25lb bags) foods and we just shed the weight and I did feel better and I had grains but they were soaked and sprouted.

We now live on a limited income so food does have to stretch so we are bad and eat grains. I have tried to convince Ed to at least go paleo but as of yet not much luck on that front though he did mention saving up to buy a gluten free bread machine system to at least get rid of the gliadin that was created in wheat starting in the 60s. We are growing Moringa trees in the shed to get a jump on spring planting since that is supposed to be a super food and because we do get cold in the panhandle of Florida we can only grow it as an annual here.

Well I have bored you enough with my saga but I thought I would add that I have back and hip issues with pain that radiates down to my feet but it does come and go. I was taught to do trigger release on myself and that seems to help the most it is just difficult when it is your entire body versus an arm or something. Oh I do also take Low dose Naltrexone that I make from 50mg tablets. I dose 3mg a night. I did not think it was doing much but when I stopped for a couple of weeks I got very depressed and lots of pain returned so I am back on it. It is very cheap when ordered from India like $60 for a year. And we do drink lots and lots of water kefir drinks but the kefir grains eat the sugar to produce the good probiotics.

One last thing. Can you do the msm will doing beck protocol since Ed is in the process of make the homemade machines.

Linda<

Hi Linda:

Good question. So do I have a mouth full of fillings. No doubt that's a very bad thing and MSM will, for sure, grab some of the mercury. At this point, we'd need a comprehensive study to know what the best action is... Swallow or spit it out. The results that I report are based on swallowing though. So I can't speak to the second.

As to your difficulty with taking it... My wife sounded like you till she realized that the results beat the heath problems... by Miles. Then she got over even thinking about it and "loves MSM". She woofs it down like candy... With dedication and purpose.

You first have to overcome your deficiency, chances are that ½ cup/ day will get you there in a couple of months. I'll send you an upgraded addition to the book.

Fibro? No refined sugar & low Carb... No grains. Learn about Duncan Crowe' s undenatured whey.

These are all dietary issues that you relate... But most things are.

Stay with it & keep your sense of humor. It's always a mind/ body solution. So don't discount this last part.

Take Care, Huuman

Hi Linda:

Missed answering this initially. Bob Beck addressed this on his videos to some degree. MSM is an all natural food, I believe that he would agree that there are no conflicts. Homemade machines seem to do lot of good judging from feedback over the years. My friend's camera flash does some nice tricks too.

You are certainly a well-entrenched alternative healer. Please let me know when you get rid of the back &hip issues. They should pass away quickly.
Try making Ed's Ice Cream per my recipe using the fruits only as sweeteners. & keep him off the bread as much s possible.

Regards, Huuman

>I scanned Duncan's page but saw no recommendations on dosages for Inulin. Does one just follow the instructions on the container?

-Alan<

Trush/ Mouth Candida
Hi Alan:

Inulin is kinda tricky for some people, but it can work miracle for most of us in terms of maxing out the digestive flora. I have not heard anything definitive on dosages; but please realize that it occurs in most veggies, so it is natural and dosage should not be critical. I usually have taken a tsp./ day (good or bad?). Problem with flora is that there is no way to know how you are doing till you get bad reports from your digestive system and it may even be too late to control the bad outcome... like a candida outgrowth, There is no pain ID like with joints and MSM.

And its not just gut flora that can be a problem. Beneficial Mouth Flora keep down the candida... found that out the hard way. You don't want that to happen. Thrush is devastating stuff once it gets a toe hold and it's a bear to beat. It probably took me two weeks of shear pain to get past that one. Boric acid appears to be the right solution for that one, but I did not know about it then. Inulin may help there too, but I have no knowledge of oral use other than just swallowing some lose. It does taste good, however, and it is a mild sweetener.

The problem for some is that inulin can cause a growth in bacteria that cause gas... yuk. No first-hand experience with that one though... guess I didn't need that.

So if you want to know much more, you are either on your own or you have to address it with Duncan... he is the guru and this, whey and slew of supplements that he sells. He both a fountain of knowledge and accessible, so shoot him an email or just call him.

Regards, Huuman

>Can I ask the group, if anyone has found a way to take MSM and mask the taste, trying to get the wife to take it is a no-no so far.
Come on folks Any ideas? Help me with this.

Peter<

Getting the Most from MSM
Hi Peter:

You can ask, but the question is, what do you want.. Maximum benefits or good taste? If you follow the protocol and swish it around in your mouth, the taste buds lose sensitivity due to the methyl component. The longer you hold it, the more benefit you get.

One person asked if she could eventually just spit it out. That is a good question and the answer is maybe... we just don't know. Basically all that you do know is what has worked for me and the people who have followed what I an now calling the *HL* MSM protocol in my book.

Many have tried putting MSM in juices for years with far worse results than I have gotten and heard from others using this protocol, but its your body. If you come up with something that is life changing, let me know, I am all ears.

I have been working on this for fifteen years and I know that, properly followed, it really will fix your joints up in a couple of months, no problem. And to beat that, you are likely to have no heart or vascular problems or contract cancer... plus, it has healed MS, I am told. The bad taste is a small price to pay IMHO for such outstanding results... plus just about everyone gets past this eventually.

Add a Low Carb Diet (LCD) and Undenatured Whey (UWPP)and you have what I am calling the AHP or Anti-Aging & Health program... and from what I am getting, this will change your entire life to the better.

Regards, Huuman

The Boric Acid Conspiracy by Walter Last
Hi Alan:

You may want to read http://www.health-science-spirit.com/borax.htm

I used it as a supplement for a month of two, but went to Swanson's Triple Boron because the wife would not touch it.

From what I found, I could do a teaspoon of it without repercussions but that is as far as I went with it. So here is what I'd do. Put a teaspoon of it in your mouth and swish it around for several minutes, then spit it out. I'd do it as often as I could find time to do it. The stuff is well known as a cure for vaginal thrush, so it should be totally benign orally.

I ended up "getting" it with a reduced laundry bleach solution, but I was desperate. It was killing me. It hurt like hell, but it cleared it out pronto.

Regards, Huuman

>Hi Jim. Thanks for the response. I'll start slow with inulin and hopefully avoid the gas.

I'm curious about your comments with thrush and boric acid. I believe I have thrush/candida and it has been a bit bad lately (eating too many holiday treats, perhaps?). How do you use the boric acid?

AHP Theory Explained Concisely
>Stephan wrote (snipped):

But while MMS may help in arthritis, it does not represent a cure, since arthritis most probable stems from gum disease or ~cavitation" (old tooth extraction cites) or root canals, essentially bad dentistry. The arthritis would surely come back when using MMS were stopped.<
(note: this discussion topic was MMS not MSM... see warning below)

Hi Stephan:

Arthritis can be easily cured with the *HL* MSM (High Level MethlySulfonylMethane) Protocol (that has already been proven first-hand), so your assumption has to be a simplification of a very complex process. However, I agree that arthritis certainly could begin as you suggest... or
with just about any kind of infection. But a gum infection is often chronic and low level, so would present a huge problem in my estimation. Also, I had juvenile arthritis and it would not be explained by your theory, but it could be explained by the following:

Here is my simple take (my mind does not do complex) as result of my research on liver function: The liver must not be over stressed or it fails in certain areas in its job of basically running the body (as the most complex of all chemical plants on earth). When you eat too

much protein,
fat, carbohydrates, etc. or have given the liver any job that stresses it such as a long term infection (of the gums, using your example), it goes into "life saving mode" (think of the guys sealed off in a submarine when it is struck by a charge and compartments are closed). In this mode, certain
of its functions are bypassed. In the case of arthritis or joints, it simply contributes less to their health and we get pain and joint failure. For this reason, I see joint pain as a very good marker for liver stress.

Liver pain, itself, would be another warning of liver stress. It is interesting that we would be alarmed and go to the hospital with chest pains, but not when we get liver pains it is passed off as not being significant. This is because the liver dies slowly and the heart stops suddenly, but both can be terminal.

My conclusion is that when provided with an adequate supply of sulfur the liver is relieved of stress, therefore, arthritis cannot occur. My take is that this is the balancing game that people NOT on *HL* MSM play and most lose.

The same is true for the protein, fat, carbohydrate part of the liver balance. If you go for higher protein levels in your diet, you must reduce some other part in order to eliminate excess liver stress. Actually, of these three, protein produces net lower stress, because the body can change
it to glycine which neutralizes toxins and relieves liver stress (a fortunate circle) . In any event, if you are going to be using lots of undenatured whey in order to increase your HGH and slow down the aging process, you must reduce either your Carb or fat intake. In our case, we chose to reduce carbs, so it all works just fine... The AHP (Anti-Aging & Health Program) is thus a successful interactive program.

Basically, my conclusion is that virtually every disease (heart and cardiovascular disease included), is related to liver imbalance and liver stress. This is why I am confident that the AHP, as introduced ***here***, will significantly increase life spans and overall wellness and health will
reach levels never previously possible in the history of humans.

So how could MMS (please do not confuse MMS with MSM people, they are very different things) help relieve arthritis? I have no first-hand proof that it does, but it makes some sense that it would if it reduces it reduces infection and allows the liver the freedom to do its full job.

Regards, Huuman

1/ 12/13 ***This supplement should be particularly interesting to those with any heath issues related to the immune system or diabetes. You could spend days reading Duncan's thorough***
response (Which is part of the reason that most of us love this guy):

>Beta-1,3/1,6-glucan

Has anyone heard of this? some pretty strong claims are being made mike<

Just one of a group of beneficial sulfated polysaccharides to discuss.

A brief orientation of several polysaccharides appears on this product page:
http://www.skinactives.com/Glycan-7-Booster.html

I see polysaccharide therapy as an oral one myself but there is some evidence of cutaneous absorption.

Short chain sugars as well as sulfated polysaccharides and fucoidans are incorporated into my "glyconutrient pudding".
http://www.remedyspot.com/showthread.php/1199321-Duncan-Crow-s-Glyconutrient-Pudding-Recipe>

Certainly a pinch of other polysaccharides could improve the batch. More on fucoidans:
http://members.shaw.ca/duncancrow/fucoidans-references.html

Solid references for each major category of polysaccharide exist. Beta-1,3/1,6-glucan is insoluble and occurs in baker's and brewer's yeast, which is part of many probiotic formulas for bowel health.

The effects of beta-glucan on human immune and cancer cells:

" In animal studies, after oral administration, the specific backbone 1-->3 linear beta-glycosidic chain of beta-glucans cannot be digested. Most beta-glucans enter the proximal small intestine and some are captured by the macrophages. They are internalized and fragmented within the cells, then transported by the macrophages to the marrow and endothelial reticular system. The small beta-glucans fragments are eventually released by the macrophages and taken up by other immune cells leading to various immune responses.

However, beta-glucans of different sizes and branching patterns may have significantly variable immune potency. Careful selection of appropriate beta-glucans is essential if we wish to investigate the effects of beta-glucans clinically. "
http://www.ncbi.nlm.nih.gov/pubmed/19515245

Resource on beta 1,3 glucans:
http://www.ultraearth.com/products/79-extant-research-on-beta-1-3-glucans.html>

all good,
Duncan

Cholesterol is the Wrong Focus... But it has Merit
(a recent change in m own opinion following my research on this book)

01/17/13 >Joel wrote:
> Are there any new studies on the relationship of consuming coconut oil and cholesterol level?

>>Tanstafl wrote
>>Hi Joe,
>>The best thing you could do for yourself with respect to cholesterol is - just forget about/ignore it... Just focus on eating as healthy as you can (based on your own research, not what the mainstream medical community dictates)... Also - one of the reasons you don't see much explicit coconut related posts now is because pretty much everything has been hashed and rehashed over the years to the point there is nothing new to be added... Check/search the archives and you'll find answers to all of your question<<

>>> 1/18/13 "Jim" huuman60

THE BIG DEAL. What Happens When the Liver is Overloaded
Hi Joe:

I agree w/ Tans on this..>but< this HDL/LDL Ratio is apparently useful as a measure of Liver Stress. And maintaining low liver stress determines just about anything else that occurs in your entire body. The Liver not only manufactures cholesterol. It influences this HDL/LDL
ratio by at least 75% according to several sources that I read. So a healthy diet is really about maintaining a very well nourished, lean liver that is on the top of its game. There are other markers for liver stress: Common ones are: arthritis pain, joint pain, gall stone pain,

kidney stone pain, diabetes problems... the list goes on. Some of the less common ones are Cancer, Heart Attack, MS. You know... it's the stuff that takes that out the entire organism and the very thing that the liver is working so hard to save.

The liver is a baseball player running around the field with a four-foot glove trying to make the plays that others can't... because we commonly hit it with really SAD stuff. You know, lots of carbs, nasty transfats & refined corn sugars that deregulate it and make its game ten times
harder. Once caught, it chases the stuff down with a six pack of Coars Light... Does this make sense to you? So if this organism is lucky, it struggles to keep it going till it is, maybe, sixty (at the 7th inning stretch. Then it hangs on by its teeth till eighty (game lost). That means twenty years of unnecessary chronic pain passed on to you, the manager, till its heart gets its marching orders and explodes. **What a game.**

Since the liver is so self-protective, it is slow to show the results of stress, but when it is stressed, other organs begin to fail... or maybe, if you are lucky, it just passes them on in the form of skin
cancer... as it attempts to throw off a toxic overload that is beyond its capacity to clean... and saves the organism from being rained out. Remember, this organism really wants to finish the game with a win, despite what is thrown at it!

On the plus side, the liver is in one way made very much like a biological, modern automatic transmission. That is, it has many fail safe modes. However, once all of those modes are broken due to stress, it can't even do its job with blood cleansing. Then, a major organ like the heart becomes the loser. As you know, dumping crud in the circulatory system can result in a rather sudden event... One that we are reminded every day about by the media. What they don't tell us about is the real "why"... and that is... Liver Stress. The Liver really is the manager of the organism, THE BIG DEAL. We may own the player, the organism, but we generally are handing down losing decisions. In fact, most attention, currently, is not even on the game. It is on the bat boy, the heart. This amounts to blaming a loss on him... More on him below.

I have always kinda known that the liver was THE BIG DEAL... but it really hit me as I researched some of what Duncan has been telling us for years about oxidative stress, etc. for my book. When it did, I was blown over by the truth in it. The reasons are the result of some of the
most complex chains of events in science. I read and understood them for maybe an hour... just long enough to visualize the relationships and get this concept. But architects, are generalists, we always start with diagrams and this was about following diagrams. The

details behind the
diagrams will blow you away, but the solutions are simple enough for third graders to understand.

Finally, we are not going to hear about this anytime soon, just because mainstream medicine has far too much money invested in what it sees as the basis of our problem... the heart (in my scheme, the bat boy). If it reversed itself, the first to lose would be big pharma and statin
drugs... Whoa baby!!! A Trillion dollar drug type would suddenly bite the dust. Coars, Miller, General Mills, and nearly every other multimillion dollar food company would go bust. What a fire show! That amounts to both Leagues!

No, this is going to be a slow, painful transition that is coming. Yes, it will be a game changer. If you are listening and thirty, you will get to see this occur. Even if you are eighty, and subscribe to our Anti-Aging & Health Program, this could be just a mid-game show for you. That four foot glove will come off... when we discover that the liver can actually field barehanded!
Imagine that! Everything just gets easier when stress is relieved.

So backing up, as Tans suggests, HDL/ LDL is a very poor marker for heart health. Relative to the heart, it is discussing the ratio of bats broken after the game is lost. Also, the real significance of this ratio is obviously not at all understood by the two mainstream cardiologists that I know, personally. What is more, these two are probably indicative of the level of understanding of most cardiologists in the field. They orchestrate a losing game and very carefully organizing bats so that we can strike out.

Ain't This Fun,
Huuman<<<

Mitochondrial failure
1/17/13 Posted by:
"Jason" eytonsearth3

Hi Duncan:

I absolutely agree with your HGH and whey approach. Glutathione is extremely important. I did a serious of Glutathione IV & PC IV treatments as a quick fix while starting drinking morning smoothies with undenatured whey protein years ago.

One of the primary problems with mitochondrial failure, as a syndrome, is breakdown in the reconversion process.

ATP (3 phosphates) is converted to ADP (2 phosphates) with the release of energy for work. ADP passes into the mitochondria where ATP is remanded by oxidative phosphorylation (i.e. a phosphate group is stuck on). ATP recycles approximately every 10 seconds in a normal person - if this goes slow, then the cell goes slow and so the person goes slow and clinically has poor stamina i.e. CFS." (Snip)

Dr. Stanislaw Burzynski's Cancer Treatment
Personalized targeted gene therapy 1/19/13

Much is being made of the since Dr. Burzynski's has just won his court case and it has put him in the news. Today, Mercola posted an article on it, plus two related videos on his site. articles.mercola.com/sites/articles/archive/2013/01/19/cancer-doctor-burzynski.aspx?e_cid=20130119_DNL_art_1&utm_source=dnl&utm_medium=email&utm_campaign=20130119

So what does all of this really mean to the alternative world? My answer is, probably nothing.

First off, his treatment involves a targeted gene therapy (and targeted always means drugs). These therapies include what he calls Antineoplastons which turn out to be what further boils down to altered amino acids. Some of these are in the form of altered peptides and some are altered amino acids.

First, the practitioner gets a blood sample and from that they gather a molecular profile and check out the genetic sequence. From this sequence, they determine which chemotherapy drugs to proscribe. The selected chemotherapy drugs sometimes have no relationship to the cancer type. Doing this confuses patent issues and thus we have court fights.

This war has little or nothing to do with wellness: While his treatment may initially shrink cancer tumors just as chemo always has, it also, as it always has, generally come back with a vengeance... the reason for the cancer has not gone away. Truth is, it will nearly always come back... And probably with a vengeance, because these patients are not educated about the cause. Everyone including the doctors are dead focused on effect. The patients seem better, so they quickly go back to their favorite FDA approved habits: they are still drinking six packs of Coars, eating a daily Snickers Bar to give them an energy spike, and eating plenty of fries fried in transfats. This war is about making money and lots of it by treating sick people and keeping them sick.

Duncan often speaks of a non-ribosomal peptide... Glutathione in the Yahoo Groups. His path is clear: Create a more open and available source of Glutathione. This is non-targeted and thus is an organic path to heading off the incidence of cancer initially... plus, just about every other disease known to man. Simplified, *HL* MSM and Undenatured Whey are both food source of organic sulfur and both help to avoid cancer. The bottom line here is that, "It's the Liver Stupid" as I rediscovered when researching my book. The Anti-Aging & Heath Program is a natural, direct, and permanent way to get to overall wellness. Conversely, once your liver is compromised to the point that you actually know that you have cancer, your treatment is going to have to be very intense compared to all others, as Dr. Max Gerson showed us fifty years ago. He cured many cases with just diet alone. Yes, cured. It can be done.

Cancer and heart disease are serious issues. I need not say that, except that this means a lot more than most people think. It means that your liver has actually determined that it must compromise a major organ by redistributing toxic garbage to a generally sealed site somewhere in your body. It is sending what we see as a devastating disease, Cancer, somewhere else, in order for it to continue to function, at least at some level... so Cancer is really not even a disease. It is a life saving intervention, a reaction to liver toxins. In the case of heart disease, it is giving up on a critical organ just to give you a few more years here. This is just another example of how wise your body is compared to how stupid the human mind can be.

So it boils down to this: This is just another battle in the drug war. Although the war on Cancer will be getting in our pockets, medically speaking... and probably for many years to come. You do not have to participate. No one will give you all that you must finally tailor the AHP to your own body. This is finally a spiritual adventure that you must undertake.

Regards, Huuman

For a good description of the causes of Cancer, read:
http://www.cancertutor.com/Articles/What_Causes_Cancer.html

The below dialogue started with Gary reading my pea story, as one might guess, which led to his buying MSM in Canada. Its in reverse order (as posted)(The last two were emails)
1/18/13 Hi Huuman:
I did find some at a pet food store one that sells stuff for horses and home critters as well.

They have for people as well but mixed with other ingredients and in gel caplets. I opted for the horse powder. It doesn't have the best taste in the world but it isn't as bad as some people say on the net. I put in my mouth followed by a mouth full of water swish it around till water and MSM warms up then swallow followed by more water to get rid of the taste and done.
Is there a set way of taking the MSM as before meals or after meals, time of day sort of thing?

Gary D.
I don't, but they have horses in Canada... go to a feed store and buy a bucket there. Especially feed stores have a rapid turnaround. That is the most important thing. If they sell it loose, just shovel some into a paper bag. Then find a big glass mason jar to keep it in where it can't pick up outside influences. It is really good at that.

I know that Duncan buys it for his wife, but do not know where.

Let me know what you discover. If you find something good, post it in the rooms for Canadians too.

>1/18/13 Hi Jim
> I have been trying to buy MSM I having a problem finding a Canadian supplier as US companies will not send it across the border into Canada. So far I am stumped. Health food stores have a type of knock off product in gel caps not the same from what I have read. If you know of any that will send to Canada it would be great.
>
> Gary D.
> Its in the Group Correspondence Gary, but, quickly, brand is not important. Search Amazon MSM for Horses.>

>1/10/13 Gary D wrote:
>> Hello Jim
>> Thank you for your reply.
>> Do you mind if I ask you what brand name you use for your MSM?
>> I will look around for a dealer here in Edmonton see what I can find.

>> Gary D.
>>1/ 9/ 3 No it is not... I'll explain it all.
>> I hope to visit him someday.

>>>1/9/13 1:26 PM, Gary D wrote:
>>> Hi Jim
>>> Thank you for the link to Duncan's site. I live in Edmonton Alberta I was thinking I would go out to see Duncan in BC I goggled the trip found out it would be a 1700 km trip ouch. I think emails might be easier for now.
>>> Say what is MSM is it a type of salt? What does MSM stand for?
>>> Thank you for your help
>>>
>>> Gary D

Drugs V Natural Interventions
12/ 29/ 12 Hi Gary:

That was pea, not pee. I guess you have figured that out by now. Frequent urination is just another sign of aging. Aging is not linear. This and every other heath condition will go away slowly if you get your diet right.
Read Duncan's site and follow his advice... this too will reverse.
when it comes to anti-aging. He has been a good on-line friend for about twenty years. We both read a lot and help each other stay well.
The undenatured whey protein diet and MSM, together, will do a lot for almost every health condition, but neither is a targeted cure. Except that almost all joint problems stem from a deficiency in organic sulfur.
Undenatured whey protein has some and MSM is basically all organic sulfur.
These are foods, not drugs. Drugs always have repercussions... always keep that in mind. Drugs may make things better for awhile, but they almost never cure anything.
Wellness is an attitude.
Jim

bone density loss
> Jan 24, 2013
This is so amazing! Ozonated oil helped to heal open wounds in the mouth of BRONJ patients within 10 days!
From: Dr. Frank Shallenberger [mailto:FShallenbergerMD@Letters.RealCuresLetter.com]
Sent: Thursday, January 24, 2013 8:02 AM
To: sfuelling@sbcglobal.net

Subject: These drugs destroy your bones... but this simple treatment stops the damage
http://securepubs.com/graphics/RCAlert_Masthead3.jpg>
Volume 6, Issue 4 | January 24, 2013<

These drugs destroy your bones... but this simple treatment stops the damage
This is a story about a terrible problem. One that is becoming increasingly common. And one that we can cure with a natural, easy, safe, and inexpensive solution. The problem is called bisphosphonate-related osteonecrosis of the jaw (BRONJ).<

Bone Loss (Rewritten & re-posted 5/1/13)
Hi Group:

Bone loss comes in many forms. Herein, I mention a couple, but they are all associated with the classic SADiet which the FDA seldom implicates, since they endorse their food pyramid and collect their money from the mass marketed, packaged garbage that they class as food.

The fact is that Vitamin D3 alone could turn the bone loss problem around according to the Vitamin D Council's recent report. Vitamin K2 is also helpful and the reasons are listed in my book. However, the main treatment, if you do the research, should boil down to organic sulfur (High Level MSM), the Low Carb Diet, and Undenatured Whey... These deliver in spades. I also, today, always recommend your reading, "Deep Nutrition" by Dr. Catherine Shanahan for a thorough discussion of the Low Carb Diet. If you incorporate these elements in your diet, the following discussion will be an unnecessary read, but let's continue anyway:

Ozonated oil may be a great treatment for bisphosphonate-related osteonecrosis of the jaw or BRONJ, also known simply as osteonecrosis of the jaw or ONJ. It is interesting that the FDA (on their web site) suggested that Vitamin D should be taken with any Drug treatment (along with calcium unfortunately). But at least they got it half right... and, to be fair, they may be improving, but it is too soon to say for certain.

According to the following article on the web address noted, you should focus on the two main issues to reverse bone density loss: http://www.boost-bone-density.com/2StageProtocol.html

The following two paragraphs are snipped from the above article (with my comments):

1. Improve overall absorption of nutrients into your cells. As we age our cells become toxic and we do not uptake nutrients efficiently. This, of course, is a description of a hotbed for cancer. Any sound recommendations for increasing bone density will also help prevent cancer because they will detoxify your cells and increase nutrient uptake into them. (IMHO, it all comes down to what Duncan has been telling us for years... and the bottom line is that

the Liver is where it all happens. The master anti-ox oxidant, glutathione, is a key link in this chain of events.)

2. Turn on the repair process in your body. You can start by giving your body the nutrients it needs, yet little happens because an aging body does not have enough growth hormone and other factors to repair itself. (My comment: While these are not really aging problems, per se, they are conditions that we commonly attribute to aging.) In this condition, the healing process does not turn itself on easily and does not work as rapidly as before. That is, we have neglected to pay attention to the factors that lead to a healthy liver, adequate minerals. and a healthy supply of glutathione.

They go on to recommend Magnesium and Calcium. I agree, but I recommend dietary sources for calcium (only). Even though a recent study suggests that supplementary calcium does not cause heart disease "in itself" calcium supplements definitely cause muscle contractions (read cramps). That one I have proven first-hand... So what is a heart attack but a massive cramp? Actually, the mainstream (FDA, MD's, & Food Companies) loves calcium and you get plenty of dietary supplementation just from packaged food products, so this is just another reason to avoid it if you use packaged products. My advice is: Always get your calcium from fresh greens.

Magnesium may be a laxative, in quantity, but the key to taking it is to supplement it to the maximum, then complement that supplementation with liquid transdermal magnesium salts. If dietary sources of calcium don't reverse osteoporosis, the above article suggests that you may have one of the following additional problems:

- Excess calcium intake combined with too little magnesium (common for most people).
- Sex Hormones being low or out of balance (common for women beyond menopause).
- Low, poorly functioning thyroid (common).
- Candida Overgrowth (common).

However, the Anti-Aging & Health Program, especially Undenatured Whey and MSM along with the Low Carb Diet will not only reverse osteoporosis, they will reverse many other problems. So there is an easy long-term solution that requires no special treatment like Ozonated oil (which I am not discrediting) and will not lead to complications.

Now, on to the drugs: Have you listened to the commercials for Boniva, for instance, and heard the warnings? Does anyone believe them? Are you aware of the problems of all drugs designed to alleviate the problems associated with osteoporosis: Actonel, Fosamax, Zometa, Boniva, or Aredia (oral bisphosphonates)? Most here know that drugs rarely, if ever, cure anything. The above drugs are no exception. In fact, in this case, they exacerbate the problem as described below.

Esophageal adverse experiences, such as esophagitis, are occasionally associated with bleeding and rarely followed by esophageal stricture or perforation. ONJ (mentioned above) is a condition in which the bones of the jaw disintegrate. Severe, and occasionally, incapacitating bone, joint, and/or muscle pain are also experienced.

To be fair, reportedly, 94% of the ONJ cases occur in cancer treated cases. But they do occur, often enough, with the above oral bisphosphonate drugs used in osteoporosis treatment. Since few doctor are trained in nutrition, they seldom are aware of the cause and cure of osteoporosis (unfortunately), so they will generally suggest taking these drugs. I suggest that if your doctor does, it is time to consider looking for one who understands the word "cure." It is well know that bisphosphonates do not create natural bone and though it looks thick in X-Rays, the bone is "structurally unsound" (in the words of an architect).

Kind Regards, Huuman
Regards,
Huuman

MSM Success/ Predictions
1/24/13 Hi Gary:

Sorry to hear about the groin pull. They can be very painful. Thanks for posting this for the others in Canada who may have probs getting the MSM.

And you are only at 1Tblspn/ day, so far!

Story: I have a close friend who weighs maybe 375 lbs, who just crossed the 2 moss mark. He is really pretty compact for such a big guy, but you can imagine the stresses on his knees and they had paid the price. He stopped by yesterday to tell me that he had NO pain after having been unable to sleep well (because his knees were going) for almost eight years. I tell you this because: You ARE going to get there.

The slight headache tells you that you had some toxins that needed to be dealt with and your liver is coming along. Wish that I could have reached you before the knee replacements, but so be it. My thought is that this AHP with help preserve even the crap they stick in us, but that one has to be tested. Furthermore, I predict that your back will heal fully and your "urgency" will be a distant memory by summer's end. Not sure about the carpel tunnel ops, but they will improve and the spurs will go away (mine did, and quickly).

Kind Regards, Huuman

>Hi I am new to this group I have chatted with Duncan and Jim of late and thought I would try MSM. Just a little history from me to start. I am 55 about 85lbs overweight, had a broken back eight years ago 3, 4, an 5 vertebrae broken in a fall. Two knee replacements twelve years ago, two carpal-tunnel operations on both hands and heal spurs to boot. I was a welder for thirty years standing on concrete all took its toll. Recently in the last year I started with a urinary problem which of course can't go a full night without having to get up and go the bathroom. I gather the problem is a swollen prostate so I am going to try MSM to remedy this and maybe my joint pains as well.

I live in Edmonton Alberta I say this because it is not easy to get products from the US into Canada. Although I found a pet store that sells MSM fairly close to me so all is well.

My results after four days of taking the MSM. I have been taking 1Tbl spoon twice daily first thing in the morning and about one hour before bed. I haven't changed my diet yet I will wait for another week before doing so. I have had a slight headache but not to bad. In the last couple of days I have noticed I can go longer between having to pee. Last night was a full five and half hours that I didn't have to get up. Also the urgency of having to go has become less than it has been. Back in the middle of December I slipped in my garage practically did the splits I pulled a groin muscle. My doc told me it would be months before it would heal. Up in till I started with the MSM it has been a real pain, literally. I feel I am up 50% of original which is far better than it was.

That is about it so far I will chime back in about two days and give an update.

Gary<

The most alkalizing substances on the Earth/ Using (bi)Carbonates,
1 14, 2013 Hi Stephen:

The salve looks very interesting; very expensive though!

Some things are counter-intuitive. For example, although MSM, as a sulphur, is acidic, it is a strong alkalizing agent in the body. It literally strips acids from soft tissues, so the pH of many people will first drop as the body cleanses.

Lemon is a fantastic alkalizing agent, acting as a blood purifier and liver cleansing agent as well.

High quality fatty acids are alkalizing even though they are acidic, because fatty acids are required to deliver nutrients intracellular.

While some people have great ideas, such as using magnesium, or using potassium, magnesium is not well adsorbed in the intestinal tract, so in order to make a big difference using magnesium alone, one will get loose stools.

Magnesium Bicarbonate is the best form of magnesium to use; about 50% is bioavailable.

When using (bi)carbonates, they should be balanced. In other words, don't just use sodium bicarb, use sodium bicarb, potassium carbonate, and magnesium bicarb.

The most alkalizing substance on the Earth that I've found to date (natural and safe, that is), is nine times roasted bamboo salt, imported from Korea. It has a pH of 11 and an ORP of about negative 600, making it a fantastically balanced alkalizing salt with significant antioxidant
properties. I purchase salts and minerals from all over the planet to study them.

Here's an article I wrote about balancing the body's minerals:

http://www.greenclays.com/earthcures/alkalizing-balancing-bodys-minerals/

It's a bit more on the advanced side of things, designed for those individuals who haven't been able to restore their metabolic state to homeostasis.

~Jason

Two Feathers Formula
1/14/2013
> Hi Jason,
>
> That is so interesting!! As I stated, I know too little about it, I only know you have to strengthen the body's buffer system but don't know how.
>
> Yes, I am eager to learn more about this important topic.
>

> By the way, the 'Two Feathers Healing Formula' has amazing alkalizing effects. It was a while >ago, I tested my saliva first thing in the morning and it was slightly acidic (6 -6.5). Then I took >one of the Two Feathers tablets, maybe it was for a stomach issue. The next morning I tested and the
> pH strip was dark green, very alkaline! First I didn't put the two together but then I realized that >it must have come from the tablet.
>
> Regards, Stephan<

research your supplements
Jan 14, 2013
Hi Duncan:

Agreed. That's why we have to research the devil out of what we supplement. One thing that we do know that the food that we get in stores will just not get us there. Its about optimization though &
thanks for your input on that.

Regards,
Huuman

>My thought is that many supplements are "too weak" to do anything. It's a common sale in fact.

>If you use Black Indian Shilajit or Mumio for alkaline minerals you'd be better off with the trace >elements.

>all good,
>Duncan<

Magnesium
Hi Jason:

As to Mg as a supplement:
I agree that magnesium is difficult to absorb. The reference that you give for Milk of Magnesia (MgCO_2)> _(MgCO3) is interesting. I have never tried a clear liquid form, so I know that I have never tried this as a supplement, but I have tried about every chelated and organic form sold and they _all_ turn out to be laxatives as you suggest. However, according to wiki, as I expected,

Magnesium Bicarbonate (MgCO_3) is also a strong laxative, so aside its 50 % absorbency, how is it a solution in itself?

My point here is that it's the Mg, alone, that sets the tolerance rate, not the CO_2 or CO_3 .So once you arrive at that point, you need another solution, and all you are doing is altering the chemistry of the laxative and changing the amount you can take in.

My solution is transdermal once I get to my tolerance rate of the chelate (I am not a fan of chelates though, mind you).

I did like reading about the Mg_CO_3 adding to the lifespan of animals... but question if it would do the same for people who achieve higher amounts of magnesium already (as I do). My reading led me to believe that you absolutely need all of the magnesium that you can get comfortably, while avoiding the laxative effect that we know is counterproductive.

I agree also that, "Lemon is a fantastic alkalizing agent, acting as a blood purifier and liver cleansing agent as well". Add the peel + lemon to Undenatured Whey Smoothies from an organically sourced lemon.. and even better. I grow lemon trees in my house for this... as house plants (mine died and I am starting a new one). Tree ripened lemons taste as good as oranges to eat and are very refreshing... same with limes I would guess but I have never tried raising them.

Kind Regards, Huuman

> While some people have great ideas, such as using magnesium, or using potassium, magnesium is >not well adsorbed in the intestinal tract, so in > order to make a big difference using magnesium >alone, one will get loose > stools. Jason

Lemons
Hi Jim!

Yes, lemons themselves are acidic; when they hit the stomach acid, all of the alkaline minerals are released... How you actually tell whether a substance is alkalizing or acidifying is by measuring the ash after its completely burned.

Here's an interesting basic article that explains lemons:
http://phbalance.wikispaces.com/Lemons+Alkaline%3F

I know a few severe alcoholics with bad liver damage that repaired their livers with lemon water alone; some had to drink the juice of five or six lemons a day.

As a balneologist and balneotherapist, I study a whole bunch of different kinds of water. Acid water is known as proton donating water... IF the right acids are used, the net biological effect is
alkalizing, providing that the body has an adequate reserve of alkaline minerals on-hand.

One of my favorite waters to experiment with is water that Jim Carr calls Hydronium water. By using a very strong calcium sulphate acid diluted into water, hydrogen atoms are actually displaced, functionally creating H3O3+ in the water... so, in this case, a hydrogen donor. http://www.altcancer.net/h3ointro.htm ...a very interesting water additive!

Most ill people that I have experience with need a lot of work restoring minerals in the body, and in order to do that, the body's biochemistry has to first be normalized. Massive infections, cancers, toxicity, they all lead to an abundance of acidic waste build up in the body.

That's one reason I love MSM so much; MSM strips all of those acids out, and it can do it so quickly that a toxic person can feel like their about to die if they start off with too high of a dose.

So I have interested individuals monitor their pH levels through saliva testing while starting MSM, and have them cut back if their saliva pH drops to around 6.0... otherwise, a lot of these types of people would freak out and stop taking it.

I myself rotate the use of my MSM; I take about ½ cup daily for one to three months, then stop taking it for awhile.

I live in a city, and city water is contaminated with a lot of stuff that they don't test for, such as xenoestrogens, which can have a negative biological impact in the parts per BILLION. We buy our drinking water; 7 stage RO filtration, ozonated; then I add back in the minerals I want, so I know exactly what's in the water. I use a Hanna PWT Meter, and a Hanna Multimeter to test the PPM and the pH before adding in the minerals.

Acids aren't bad; my reference is always dealing with individuals who have an abundance of acidic waste built up in-body.

~Jason

Lemons
1/16/2013 Jim wrote:
> Hi Jason:

>
> Lemons are much more acidic than 5.5 by assumption and I'm told that> they have a net alkaline effect> I am not sure why... makes no sense to me, but that would tell you that your readings may be all wet (pun intended).I'll have several electronic pH meters here and maybe one works. I'll
> have to check this out. My other thought is that MSM in water needs to be slightly acidic
> just to make it work.
>
> By the way, where'd you get the alkaline water... is that your tap > water? I am told that slightly acidic water is more healthy, but it should soap up well. Ours here is almost dead neutral.
>Cheers, Huuman

Dr.Terry Wahl's
1/17/13
(Jason on the Dr.Terry Wahl's protocol & why it may not work, esp for CFS sufferers)
Hi Gail:

I have a real problem with someone recommending a high fruit/berry diet for CFS sufferers. You may be able to get a temporary fix with this, but an eventual crash is likely.

I like his idea of trying to flood the body with antioxidants. After years of research, I've personally concluded that fat is a better source of energy than sugar. Pushing fast burning sugars can be catastrophic for a small but significant portion of the population.

I tried this method of recovery early on. It vastly encourages latent infections to thrive. When I started to get worse, I kicked myself, because I should know better. For some it might work well; I went in and had my metabolism tested, and found out that part of my impairment was an vastly reduced ability to burn fructose, in any form. By eating a lot of fruit, it "appears" that digestive health improves and that energy conversion improves, but this is often just a nice illusion...

It is not sustainable for a large portion of people, just like running the body on caffeine can be done...but only for so long before the adrenals start to fail!

Please don't misunderstand: I'm not anti-fruit. Even people with intolerance can eat moderate amount of fruits - I use blueberries, black berries, and sometimes strawberries in my morning smoothie... but in extreme moderation.

The cure comes when whatever caused the mitochondria failure is gone, and the little mitos are recharged and repopulated. A lot of these methods out there are ways to steal from Peter to pay

Paul; and it may work for some. After all, if you detox the body enough and heal the immune system, recovery is natural.

I absolutely agree that proper zapping/pulsing, blood electrification is a great idea.

I wish I had purchased a stand alone pulser; mine attaches to an EMEM3 device I have, is a little bulky and inconvenient to use, so I haven't used it much!

Oh, I almost forgot: I took Wahls' idea of antioxidants, and supplemented by using LOW glycemic extracts. It's more expensive to use extracts, but it's the only way that I personally could make it all work.

~Jason
:
> Hi Duncan & All,
>
> Have you or others followed Dr. Terry Wahls' protocol. According to her, mitochondria recovery is in the fruits (berries)& vegetables, lots of them as in 9 cups daily, along with cutting out dairy & wheat and eating grass fed beat and other quality protein. Following this diet, she has recovered from MS and helped others with various illnesses. Inherently, she said when one cheats, symptoms start to return and it will take 2-3 weeks to clear them when returning to the diet.
>Gail

Liposomal Vitamin C
1/22/13
Hi Jim:

Yes, I really like liposomals; they really are more effective than other forms... the downside is the price. I've used liposomal glutathione as well; it's not as good as using whey, but even if using whey, taking glutathione for a period of time can be a good addition to a detox protocol. It's also a good way to test liver and blood health; taking liposomal glutathione will pack quite a punch if cleansing is needed!

Liposomal vitamin C is excellent, especially for people who may have a low tolerance for buffered C,. ascorbic acid, etc. There are no digestive/bowel problems using it.

I think individuals should be getting at least three grams of vitamin C daily, and individuals dealing with illness, more. However, there is a fun way to test how much Vitamin C the body needs.

It's not a perfect method, and you really need a day off where you don't have to be anywhere; it's called the Vitamin C flush. Rather than type out the instructions, I'll just post a link: http://www.livestrong.com/article/498842-how-to-do-a-vitamin-c-flush/

By following the instructions (it's important to use BUFFERED vitamin C for the flush), one can determine, roughly, how much Vitamin C the body needs; i.e. how depleted the body is or is not. If using liposomal as the supplement, one can probably supplement at about 1/4 the amount that the
flush indicates with buffered Vit C.

Regarding the flow chart idea... I started developing detox protocols, bowel cleansing protocols and digestive healing protocols in my mid-twenties. Using edible therapeutic grade clays is a specialty, and I enjoyed helping others with these type of things. About 85% of the time, the limited methods I used worked perfectly. But in my mind, they shouldn't have been effective 85% of the time, they should have been effective 95%+ of the time. So I simply said 'fare thee well" to the individuals who now had the tools they needed, and when opportunity presented itself (other brave souls), started an in depth study of the failed cases.
We learn far more from failure than success.

Eventually, what I noticed is that nearly 100% of the time, the failures had something in common: Acidic waste build up in the soft tissues.

So, I began developing steps... In other words, do protocol A for three weeks, retest, and then if success, continue on, if failure, return, and add protocol B... if failure, return, and add protocol C.

Each step in the process, ideally, is designed to eliminate or rule out a particular problem.

The idea being to set a clear goal that is measurably achieved, allowing room for individuals to be just that: individual.

Eventually, the end goal is to have this sort of system present for every main physiological system in the body... i.e. a program for digestive health first... including assimilation (eating habits) and
elimination (detox)... a program for immune system strength (and pathogen elimination), cardiovascular health, muscular health, hormonal health, and neurological health.

Since everyone is an individual, I believe that health has to be sculpted for an individual... I've never found a one size fits all approach. Finding the right starting point, and given the right education, amazing things can be achieved.

Of course, I'm nowhere near qualified to deal with every system of the body... nobody is. The endocrine system is particularly puzzling, and while I understand more about neurological health than most, I often must consult with others. ...at some point, an individual reaches the point of halting degeneration, and begins the process of regeneration.

People can do this for themselves, at whatever level and pace they choose to.... hopefully before chronic and degenerative disease reaches the life threatening state.

I'm probably a good ten years away from publishing my next book, and I need to completely restore my own health before seriously considering doing anything on a formal level.

You're lucky to be able to grow your own plants/trees! Getting the mitos to a happy place has been confounding.

For example, I can do 100 squats/day with a 20+lb weight add-on, about thirty pull-ups with leg lifts (pulling the body up, lifting the knees the chest, then extending the legs straight out), I do about 100 back extensions daily, and 2 minute planks, with weight training about once weekly (functional weight lifting, not bulk building). Believe me, with individuals with CFS, this is not a pleasant experience; in fact, it is quite painful.
And still, I'm limited in what I can physically accomplish, such as standing still, sitting in a vehicle, etc... Sustained energy production and reconversion is not anywhere near optimal.

~Jason

1/22/2013
> Hi Jason (& Group):

> With regard to the flow chart approach (maybe we can pull some others into this):
> It is about how you approach the flow chart solution and how you make the mitochondria
>happy.
> Mainstream solutions might attempt to solve it by attacking each branch of the chart with a
>different drug. You might give some examples of what you mean to the forum, from an
>actual chart, so that we can see what how it works. It sounds very comprehensive. I am
>thinking that you could end up creating a whole new field within the wellness community by
>getting this out.

> Next, what do you know about liposomal Vitamin C? Do you use it?

>I am just now researching it after reading Mercola's stuff. Till now I have relied on getting adequate >C from citrus and especially organic lemons (as you suggested about a week ago) which are great. I >actually eat the entire fruit, using the peel in my smoothy. The peel will really jack you up like >moringa leaf does, but they are hard to come by. I am starting a new tree from seeds, so it will be >awhile, LOL.

> Regards, Huuman

> Hi Jason:

> I am about as sure as I can be about this one, my new friend. This stuff is not for the weak at heart (figuratively) . You are doing everything that I now about at least on the physical level... there is
> more, of course, but that is another story.

> Cheers, Huuman

> Hi Jim:

> Wow, thanks for the vote of confidence. :) Failure is simply not an option; for the mind that works >with raw creativity, everything in the universe is connected, and everything has
> a valid meaning and purpose. The hardest part is weeding through the often well-intentioned >but often misleading information. How one thinks is paramount to what one sees.

> With all "mystery" syndromes, chronic illness, and disorders, where one puts one's mind is critical.

> What I'm developing is a flow chart like system with key things that must be accomplished, then >simply finding ways to accomplish them, and of course, one always, always questions everything.

> The hardest part is getting the mitos back to a happy place. Perpetual tissue damage occurs whenever ATP energy re-conversion can't keep up with the demand (rapid aging).
>
> ~Jason http://www.greenclays.com/earthcures/

MSM Taste is Worth the Effects
Tue Feb 5, 2013:

Hi Craig:

You are not giving an accurate description of MSM taste... First off, it is neither bitter, sour, nor sweet. It is the strong astringent (not even a basic taste) of the methyl groups that you dislike. As I have related to everyone here many times, those methyl groups are also, basically, temporary taste bud killers. The best way to take it is to just scoop it in your mouth, follow it with water, and swish it around till it numbs your taste.

The idea is that you WANT that methyl component into your system pronto.... and this will make it happen. If my wife can do this, anyone can... she is a food connoisseur (nut if you will) from the word go.

If you do it your way... or mix it with water or juice in a container, you are losing one-half of the value of this amazing organic chemical. So what you are suggesting is a problem, unless you are not serious about this actually doing anything for you (so, why in the first place?).

As Duncan warns, bananas are high carb fruits, also... So, you are doubly messing up with this "off the wall solution."

Keep it pure and do it correctly or you are just scratching your butt as we'd have said in the Air Force (with a bit more authority).

Regards, Huuman

>Greetings.
>I have an idea that may help everybody with getting past the bitter MSM taste. Measure out
>the MSM that you want to take. Dip a banana in it. The MSM will stick to the banana. Eat it.
>Repeat until all the MSM is gone.
>Works pretty good.

>Craig<

Natural boost that works
2/8/13

Hi Mike:

Let's be clear on this. Herbs are very different from minerals and vitamins in that they all contain micro nutrients & phytonutrients that can do some very special things, But I agree with the wisdom of Dr. Joel Wallach on this one: if you really get a huge boost from an herb, it is probably because it is providing you with a trace mineral that you were lacking in your basic diet. That means you should be supplementing that mineral on a daily basis. The problem with herbs is that they may only provide that trace element for a short period of time according to where they are harvested. plants can never "make" trace elements. They can only take them from the soil. If I were you, I'd be supplementing with colloidal trace elements first and that Longefolia second or you are almost bound to lose this nice "boost" that you are reporting.

Especially, these wild herbs come from the mountains and are not cultivated. That means they can often have what you need and have been missing, but again do not depend on them for long. Suppliers switch sources in a heartbeat and you will be heart broken... so head them off.

Whey and MSM are more basic nutrients and they can't provide these. VCO might if you were very lucky, .since it actually grows from soil originally... are you getting my point?

Regards, Huuman

>Thanks to both dunc and jim on this one. I haven't been this excited in a long time. Finally
>products that actually work. On the long jack now and noticed a difference in energy.
>Something that the whey and vco >just hasn't provided. am looking forward to the other
>herbs on the site as well
>mike

Magnesium
Feb 5, 2013
What do you think of Jigsaw?
http://www.jigsawhealth.com/supplements/magnesium
Steven

Feb 5, 2013

Most magnesium supplements are worthless... The absolute best way to get it is magnesium oil, or even magnesium bicarb..
Tanstafl

Intestinal Virus
Feb 2, 2013
"Duncan Crow" duncancrow
Seems the norovirus, if that's what it is in this area, is about the nastiest thing even the resilient young'uns have ever felt. Some get diarrhea, but not me. Feels like it nearly kills you at any

rate. Lasts about 4 days. I think I got it out in 2.5 with 1.5 quarts of 40 PPM colloidal silver and the rest of the week was the complication of dehydration and the pneumonia on it.

Good thing we've been pitching semi-optimization beforehand. Makes us stronger I'm sure.

I found that warming in the hot bath relaxes the stomach so you can get some fluids to go through and not come up. Hope it helps.

all good,
Duncan

Intestinal Virus
2/2, 2013:
"Jim" huuman60
Hi All:

Here is my cure for intestinal virus:

"Put the Lime in the Coconut and eat it all up... It will relieve your
bellyache."

Love the Song & Love the Cure... Funny thing is it did not hit home till I got sick. And its been one my favorite song list for twenty years. By the way, I did not call the doctor and wake him up (as you might guess), it just came to me in my moment of need and this is one nasty virus that is going around. LOL.

Regards, Huuman

Heart Palpitations
Feb 8, 2013
Hi Anita:

A rhetorical question: If you get up anyway... how is it interfering? No, I have never gotten up specially just to take MSM... that would be over the top, LOL. I was kinda kidding Anita... this is not the best place to jest, but it's Huuman to do it. Sorry.

More seriously: I was relating to putting my full attention on this health problem surfacing as "heart palpitations." I was saying that it is pretty far along in a chain of events that begin with Liver Stress and, therefore, warrant a great deal of attention till gone... same with Cancer and CVD.

Pardon my ignorance, but what is this "organic sulfur" that you refer to and where did you buy it? I did not know it was for sale, actually. Once you wade through all of this, you will know why MSM, Whey, and Low Carb are the correct choice. They are an interlocking team. But, I'd probably just throw the "organic sulfur" in my smoothy, if I had it, and it did not taste bad. What's to lose?

Yes, methylation is one of the keys to the whole process and you must understand it fairly well to get the most out of the AHP. It's complex, but keep reading and read the online references, especially. It hit me like a ton of bricks when I first got it. But this is not "Fun with Dick & Jane" reading. It takes some real concentration to get it.

By the way, Byron Richards is about as close to getting this right as anyone out there, which is why I reference him in my book. I suggest that you subscribe too his free newsletter. He seems to be reading my mind in many respects. I am beginning to think that he is reading this CO & the BBE Yahoo Forums. But, hey, it's all fair game. The book is mine, but the info is from all over the web. When he starts writing about the AHP, we'll know.

Kind Regards, Huuman

Heart Palpitations
Feb 7, 2013

I didn't know the difference between just organic sulfur and MSM. What I've been taking just says "organic sulfur" so I guess I've been taking the wrong stuff? Tell me why it's okay to take MSM that is sold for horses. Wouldn't there be a question of its quality?

If I got up in the middle of the night to take MSM, wouldn't I be interrupting the time when my body is trying to heal and repair itself? I've been struggling with trying to sleep solidly through the night because I've read of the importance of sleep during those hours to help my liver and I feel better when I do sleep through the night without interruption.

Thank you again for the information. I hope to read more of your book this weekend. I only got to about page 26 and may have to re-read some of that (about the methylation)
Anita
Magnesium/ Heart Palpitations
Hi Anita:

Personally, and I have read reams about it. I am not going to get too much magnesium. No matter how much I try to take... I simply can't get too much. The limit is at diarrhea... and, no, you don't want that.
It depletes electrolytes quickly.

On Blood Pressure: While magnesium lowers BP when it is too high, there is no reason to believe, that I have read, that it will take it too low. Duncan and others may wish to chime in here. But, I have never seen anything that would lead me to believe it would.

Byron's formula looks good and I am sure that he has researched it, but I am also sure that there are cheaper, equally good, solutions. Nevertheless, someone has to finance Richard's, Mercola, and Jon
Barron... and it will not be me unless I suddenly become independently wealthy.

Your last point would be credible if you tested it and you indicate that you have not. This has to be a very rare event. No matter what we are discussing, there are people out there with allergies to just about anything, even MSM, the least poisonous chemical on earth according to some sources.

We arose from the sea and magnesium oil transdermal rub is basically going back to it. If you moved to the British Virgin Islands, would you refrain from swimming in the ocean because you read that someone breaks out in hives when they do? A half hour swim in the ocean every day would be a better solution , but I have to settle for a transdermal rub. These saltz include every mineral, not just magnesium. Mine are actually based on ocean water from Maine (when I can get it), combined with Zechstein saltz; but our local ocean water is a good second solution. The further north the water source, the more salt concentration (for you non-sailors).

By the way, Anita, I applaud your concerns here. Please do not take me the wrong way. All I am really saying is: Do more research, and take your time, but don't be afraid to take baby steps. I am pleased to
hear that your palpitations are receding. Also, I predict that they will be going away, forever, within three months. You are on the right track.

Kind Regards, Huuman

Heart Palpitations
>The reason I don't take more than that is because I know magnesium can lower blood pressure and >mine already runs low. I read that a person needs 3 mg to 4 mg per pound of body weight. I weigh >120 pounds so 400 mg is almost in the middle of 360 and 480 mg and I feel

safe there.
>My palpitations have been so much better in the last week or so and my heart feels stronger. The >only changes that I can think of are taking CoQ10 and not eating after 6:00 p.m. It could be one of >those or maybe a shift in my hormones. I guess time will tell.

>That may be true but I think the magnesium I'm taking is high quality. It's from Wellness Resources >(Byron Richards). The ones I take in the morning have malic acid and some B vitamins in them (for >energy), and the magnesium is 60% malate, 35% diglycinate, and 15% carbonate. The ones I take at >night are for sleep and >have vitamin C in them. The magnesium in them is from diglycinate, ascorbate and Kreb's nutrients.

>Besides, I read some reviews about magnesium oil from people who broke out in some kind of skin eruptions wherever they applied it to their skin, so that kind of discouraged me from trying it.
>Anita

Heart Palpitations
Feb 2, 2013

Thanks for the info. I also have digestive issues and recently started making my own sauerkraut and eating it raw with my evening meals. I was starting to feel better overall but ran out of it, and the next batch I made didn't ferment properly or something because it tasted horrible so I've been without it for a while now. I just made some more today but will have to wait for a week to eat it. I've been trying the gluten-free thing too for a while but sometimes slip a little if I eat out. I think I need to get back to low carbs again. My palpitations seem less since I restarted coenzyme Q10 but it's probably too soon to tell for sure. I think hormones could be causing a lot of my problems but am curious to see if getting my liver in good shape (like Jim recommends) will help with the hormones too.

Anita

Heart Palpitations
February 2, 2013

I do usually post on here but been reading this and thought I should post. I to have mitral valve and digestive issues. The mitral valve gets worse when stomach is irritated and it irritates my vagus nerve. I had went to a business meeting last week for the health and wellness company I am with and they were talking about Mitral valve a women had said she started taking GLA and fish oil and her palps went away. Well I had been having a bad bought of them so I tried it and it has been working great. I also take CO q 10 and daily coconut oil.

Amy

Heart Palpitations

Feb 2, 2013

Frederick Patenaude has an interesting but I think trollish remark about taking no oil if one has candida. Anyone listen to this fool? I understand VCO and MCT oils are proven to curb candida and a lot of people find it core in an anti-candida diet. So whazzup with the man? I've never heard of him but I lodged an appropriate response on his forum anyway, with my customary brutal frankness ;)

all good,
Duncan

Heart Palpitations

Jan 30, 2013Hi Carol:

I agree and how would he know, really? Certainly, you have some first-hand experience that contradicts what sounds like his "informed anecdotal level" information.

I'd be all over the coconut oil and I'd be looking for complementary additions to it.

Question: Are you doing Low Carb, Whey and MSM also? How about Ubiquinol and D3?

Regards, Huuman

>There's no rhyme or reason to them, no precipitating or alleviating factors. They don't happen every day. EXCEPT: I've noticed that when I get some coconut oil in, I don't have them. Of course, he thinks it's a coincidence, but I'm like Leroy Jethro Gibbs - there's no such thing as a coincidence! haha!<

Vit D2 V D3
Jan 30, 2013 2:02

D3 is the real stuff.... & D2 is artificial.

Mainstream hospitals commonly use D2 Does that surprise you? it should not. When it comes to diet, they still think that jell-O is a staple food.

I have no probs with cod liver oil no matter what you use it with, butD2 is a no, no.

Jim

>Why is vitamin D3 recommended over D2? Does this mean that Green Pastures Cod Liver Oil with the D2 is not so good?<

Heart V Liver Probs
Jan 30, 2013
Hi Osh:

I agree that*/chronic inflammation /*is the problem, but I doubt that their heavy handed technical solution is the answer. I also disagree that atherosclerosis the result of this problem. I think that its all about the liver as I say in my book. The gerontologists are stuck on fifty year old solutions that just do not make sense when you look at the overall solution.

When you go back to what Duncan has been telling us for years about cellular aging and the Glutathione Conjugation Pathway, what they are saying makes very little sense, but it pays the bills at the major institutions doing the studies. IMHO, its all about oxidative stress and the liver is the basic organ of life, not the heart.

Regards, Huuman

RE: this article
http://www.lef.org/magazine/mag2013/feb2013_otc_01.htm?source=search&key=telomere%2 0lengthening%20treatment>

Adding Coffee to Whey
Jan 17, 2013

No Tans, you didn't:

It is your choice, but I do it daily... sometimes several times using decaf... and it works fine. Its about how you go about it. I am not a cave man even though I may somewhat approximate their diet.

It's discussed in my book: Mix a lot of pure cream, A cap of NOW colloidal minerals, Some liquid iodine, a teaspoon of Moringa Leaf (which is also sensitive) and then pour in coffee from another cup that is now "not so hot." The idea is to keep temps moderated. I do not use a thermometer, but I have a nice electronic one if I wanted to.

Personally, I am not interested in any piping hot drinks or even ice cold drinks, ever. 130 degrees is about right for a hot drink. I do it with Moringa Tea also. It does not take a rocket scientist to know when you have destroyed whey. I did it maybe once or twice, but you know immediately. I drank those too, but they are messy. Its like drinking
clumps of butter.

Regards,
Huuman

Adding Coffee to Whey
2012-02-19
> I suggested mixing Whey into a coffee/ heavy cream mix. The Whey cream, if heavy enough and mixed well cools the coffee down prior to adding the whey.

I guess you typo'd something here, or I'm just missing something (I haven't read this entire thread, so maybe that is it)?

How can the whey/cream mix do the cooling of the coffee *prior* to 'adding the whey'?

I would never add something as fragile as whey protein *or* heavy cream to *hot* coffee... regardless of how effective it is at cooling the coffee, the damage will be done.<
Tanstaafl

Skin Care
Jan 17, 2013
Hi Judy:

Just read yesterday read in the D3 Council Newsletter that liquid D3 was good for this.
I'd try mixing it with CC Oil & give it try. You might want to look further & see if you can even improve it with some other things, but CC Oil is a great start.

Regards,
Huuman

CO Flavor
> Jan 17, 2013
"Duncan Crow" duncancrow

Most of the benefits of coconut oil apply to refined-bleached-deodorized RBD as well as VCO. So, it's one of the right kinds but purists often want it raw :)

We don't use coconut oil much because of the flavor but when it's deodorized it's a good bland cooking oil. At home we use MCT oil though because it pours at any temperature and also tastes bland.

all good,
Duncan<

>>Has anyone tried coconut oil for dry, cracked hands with good results, and did you use it orally, topically or both?? My daughter has this problem, and hasn't had much relief despite using lotions and drinking a lot of water. It gets so bad, her knuckles will crack and bleed.

Any ideas what might cause this, and other suggestions?
Thanks in advance!!!

Judy<<

There are No Golden Years
Jan 17, 2013

I agree with Duncan, except I don't particularly buy into the golden years concept. I am thinking that if we really have this down, 80 is a young whipper snapper and maybe not even middle aged yet. But this is a long term test. So far, it's interesting, though.

I'll take whatever I am given... just dream anyway,
Huuman

(snipped from Duncs post)
> Hi Carol, when it comes right down to it what we want for our own health in our "golden years" is a lust for life coz that's fun and interesting, a bit of energy to back it up coz we can't actually do much about it otherwise, a decent immune response or we won't last, and a degree of mental focus like a healthy 30-year old has or as we age we might forget where to lick. Not too hard to do with supplements :)<

> all good,
> Duncan<

Supplement Brands
Jan 17, 2013

Hi Paul:

Why use Mercola's brand? Are you trying to help finance his web page? As Duncan says, his staff is messing up quite a bit now and you don't get his first-hand advice... that is bad. I do like his videos though. I'd rather read Wiki for more mainstream opinions and background science and Dr. David Williams' old newsletters for supp advice. He does his homework.

So far, I have never seen where Mercola's brand has a "hook." Often, the brands that Duncan refs do, though. In every case, I have found that Mercola lags behind Dr. David Williams in his advice. Sometimes they lag by as much as a year behind. With astaxanthin, he was about five years behind as I recall. Also, www.healthtruthrevealed.com beats him to the punch. But, as I said yesterday, you have to be ready for their proselytizing. Both Williams and Crusador sell more intelligently designed supps in my experience and I liked Dunc's ref site yesterday.

Finally, if you want stuff that has been out for awhile, why not use Swanson Vitamins and save yourself some bucks? They do a pretty good job for me so far... and others seem to agree

Regards, Huuman

>Use Mercola brand
>Sent from my iPhone

Anti-aging and Herbals
Jan 19, 2013:
"Duncan Crow" duncancrow
Jim, men who are interested in anti-aging and herbal support should look at the blend EHP from http://therootofthematter.net

The EHP ingredients list contains a bunch of herbs we'll recognize (and several we probably don't). Maca, Rhodiola Rosea, Mucuna Pruriens, Epimedium grandiflorum, Fo-ti, various ginsengs, Tribulus Terrestris, Wolfberry, Schisandra, Saw Palmetto, Cordyceps, Wild Yam, Astragalus etc...
http://therootofthematter.net/cgi-bin/itsmy/go.exe?page=4&domain=12&webdir=therootofthematter

Although there is some attempt to increase testosterone in this blend, it also increases other androgens and is a general tonic, so it seems to me it might be a decent support formula.

I bought 3 bottles of shilajit and six of EHP, so time will tell now ;)

all good,
Duncan

Elevate Dopamine
Feb 6, 2013
Thanks to both dunc and jim on this one. I have not been this excited in a long time. Finally products that actually work .On the long jack now and noticed a difference in energy. Something that the whey and vco just has not provided. am looking forward to the other herbs on the site as well
mike

Elevate Dopamine /Mucuna Pruriens/ Maca
Jan 17, 2013 6:54
Hi Olushola & all,

I take Mucuna Pruriens and several other herbs just because of the research that I have done on them. It is fairly mild in taste and rather neutral. My study suggests that it's a good choice for some hidden and rather and not well understood chemistry. Others in this category are Cordyceps Sinensis and Maca that I take. There are plenty more, but you must do your own investigation.
Duncan just gave you some great sites yesterday.

I agree with what Duncan posted last week. Essentially, we must pick and chose on these and some just are not worth the effort. These herbs are generally not particularly life changing, but the research indicated that they are worth considering. So they are not likely going to turn your life around, but when combined with the /High Level/ MSM, Duncan's undenatured Whey, and a Low Carbon Diet, and, finally, spiritual exercises that suite you, you are most likely going to be in the driver seat here on earth for a long, long time baring an accident.

Back to herbs, my favorite all time herb, so far, is Moringa Tree Leaf (and Tea) simply because it contains a complete protein... which is unique for a plant and tastes good. It goes well in my morning coffee/ whey protein drank and It actually has a tart taste that "improves"; the

taste of most things like whey shakes and my whey ice cream (recipes are in my book).
In Wellness, Regards, Huuman

Jan 16, 2013
"Jim" huuman60
Hi Duncan:

I bookmarked it. When I run outta Maca. I'll buy them. I am also running out of room, LOL.

Huuman

>The company I get Shilajit from also sells Rhodiola and gelatinized Maca:

http://therootofthematter.net/cgi-
bin/itsmy/go.exe?page=19&domain=12&webdir=therootofthematter<

Jan 16, 2013:
"Duncan Crow" duncancrow

Elevate Dopamine
Hi Carol,
When it comes right down to it what we want for our own health in our "golden years" is a lust for life coz that's fun and interesting, a bit of energy to back it up . We can't actually do much about it otherwise, a decent immune response or we won't last, and a degree of mental focus like a healthy 30-year old has or as we age we might forget where to lick. Not too hard to do with supplements :)

Mucuna Pruriens that was just mentioned elevates dopamine, a feel good and healthy hormone that helps one remember where to lick and much more. I would caution that dopamine and the dopamine drug agonists have negative side effects and if the Mucuna (the herb) doesn't work to your satisfaction, selegiline (the drug)at least is guaranteed to increase dopamine without harm or side effects. Dopamine increase is the answer to the Zoloft/Paxil bunch of drugs that sometimes keep a person disabled, unproductive and on the couch.

There's a lot of Rhodiola info out there that wasn't available in 2004 and it looks like one of the winners.

all good,
Duncan

> Interesting, Duncan! This sounds like something I need to try - for a number of reasons. Thanks for the info!
>
> Carol

Pregnenolone

Jan 15, 2013"Jim" huuman60
Pregnenolone works for me, Duncan. And I agree that a healthy sex drive is an important element in Anti-Aging and measurement of youth. We are one organism and all parts must function well in order to be considered a successful whole.

Regards, Huuman

>Pregnenolone is now the topic of several anti-aging and healthy sex books. I just put down a decent DHEA book and looked for a compendium of current knowledge about pregnenolone. I've been taking preg for more than two years with my HGH-triggering amino acids with good results, and details from people who review whole studies will flesh out my own knowledge gained through the study abstracts, personal experience, and that of associates.

The Victoria Dolby ebook on Pregnenolone is her whole hardcopy book, more than 450 pages and just over $3 in cost. Yippee!, the free Kindle PC install was painless and the book so far is worth the effort for anyone interested in the details behind a few anti-aging principles.

As an alternative it seems the ebooks are cheaper than hard copy and another one, Pregnenolone - Your #1 Sex Hormone, by Susan M. Lark M.D. seems it might be a good one too:
http://www.wwpublishing.com/Pregnenolone_Book_p/pregnenolone.htm
Here's Victoria Dolby's Pregnenolone book: http://www.amazon.ca/Pregnenolone-Victoria-Dolby/dp/087983885X

all good,
Duncan<

Herbal Improvements May Not Last
Fri Feb 8, 2013

Hi Mike:

Let's be clear on this. Herbs are very different from minerals and vitamins in that they all contain micro nutrients & phytonutrients that can do some very special things, But I agree with the wisdom of Dr. Joel Wallach on this one: if you really get a huge boost from an herb, it is probably because it is providing you with a trace mineral that you were lacking in your basic diet. That means you should be supplementing that mineral on a daily basis. The problem with herbs is that they may only provide that trace element for a short period of time according to where they are harvested. plants can never "make" trace elements. They can only take them from the soil. If I were you, I'd be supplementing with colloidal trace elements first and that Longefolia second or you are almost bound to lose this nice "boost" that you are reporting.

Especially, these wild herbs come from the mountains and are not cultivated. That means they can often have what you need and have been missing, but again do not depend on them for long. Suppliers switch sources in a heartbeat and you will be heart broken... so head them off.

Whey and MSM are more basic nutrients and they can't provide these. VCO might if you were very lucky since it actually grows from soil originally... are you getting my point?

Regards, Huuman

>Thanks to both dunc and jim on this one. I haven't been this excited in a long time. Finally products >that actually work. On the long jack now and noticed a difference in energy. Something that the >whey and vco just hasn't provided. am looking forward to the other herbs on the site as well
>Mike<

Herbal Improvements May Not Last (Clarification)
2 13/13
Hi Mike:

I am not an herbalist either, but I have taken Longjack (Eurycoma Longefolia) for quite a long time and never noticed anything, really. But then I have no issues. Nevertheless, I take that and many others just in case... and we know that each one has valuable, unrepeatable phytonutrients.

As I said before, herbs can only do so much and when they really move you as you have reported twice here, I believe that you should get down to finding out why. The reason is that an herb is only as good as the soil that it is grown in and suppliers are likely to change their sources. You also reported having used another and that it is failing you now. That is the rub. You likely did not change, the herb did. The point is, unless you raise it, you can't control what an herb contains to any degree even if it is picked wild.

If it really is a phytonutrient that is giving you this boost, fine and dandy. However, I am inclined to believe it is not. I asked you before if you were taking a trace colloidal mineral mix and got no answer. If you are not, you should be. Many of us need more of a particular trace element in our diets.

My son had that problem. This is where genetics actually get into the picture (and genetics are, generally, way overrated). He was hyperactive till he got that mineral. We have no idea which one it is, but we know that it is there and we also know that any good colloidal source will always contain it... problem solved. That is, there is no issue with where it is grown, since it is not. Therefore, the missing trace mineral will always be available to him. You can buy Now Brand Colloidals for about $10/ quart... $.35/ day... Cheap insurance!

.Next, your boost may actually come from getting one of the more primary minerals into your system from Longjack, but this would be less likely unless your diet is really, really bad... and you are here in this group, so what are the chances? Also, chances are that you are just not taking that much... what two caps/day?

We just discussed those (more major) ones that are often lacking yesterday, but: Chromium, Magnesium, Silicon, Potassium, Sulfur, Iodine, Boron, Zinc Vanadium, and Selenium are the key ones for those who missed it. Most of us get plenty of copper and the other major minerals. And for men, we generally get too much iron, so there is never likely a need to supplement there. But the most important ones here are: Sulfur, Magnesium, Iodine, and Selenium, each for very different reasons. Our Group Archives are full of the reasons why these are important as are the Beck BE Group's.

Kind Regards, Jim

>Good on ya Mike. I'm not a herbalist but when I saw data on the tonic herbs it piqued my anti-aging >>and vitality interests.

>I've been taking the EHP blend for only a few days. Nothing to report but >tonics sometimes take >>awhile and I've been sick recently so my subjectives are a bit off and I need a turbo button.

>all good,
>Duncan

>>Dunc
>>Received the other products mentioned earlier from the root of the matter site today. a little >>background first. Had been taking vco and whey >>for some time no noticeable affect. Two weeks >>ago added longjack from the same sit, some effect but not much. took the regime today along with >>the above and within and hour the only way I can describe it is WOW. Talk >>about turbo charge. >>very , very happy from this camper. Finally after a 30 year journey something works(big smile)
>>mike

Static Cling
Hi All:

Jon Barron's newsletter (newsletter@jonbarron.org) today hits on a topic that concerns me. He ends it with:

"A study at the University of Reading in England in 2009 found that chemicals called parabens used in many commercial deodorants were present in 99 percent of breast cancer tissues taken from women being treated for the disease. Parabens are estrogen-mimicking xenoestrogens that are also used in a number of other grooming products, so it's nearly impossible to identify those in deodorants alone as the major contributor, but it is certainly worth reading labels and trying to choose products labeled paraben-free."

What concerns me is that our failed government policies (FDA) do not even address the role that the skin plays in taking in these dangerous chemicals used commonly in cosmetics (transdermal migration). Unless you happen to have read Dr. Mark Sircus' book, " Trandermal Magnesium," it would not enter your mind that the skin is so susceptible to the transdermal take-up of virtually any chemical, but it is a wonderful delivery system for magnesium... and a horrible one for toxic chemicals.

Actually, parabens are probably far less toxic than formaldehyde discussed below because they are thought to (Wiki Quote): "Studies on the acute, subchronic, and chronic effects in rodents indicate that parabens are practically non-toxic. Parabens are rapidly absorbed, metabolized, and excreted. " But they are making lots of money with them and there are commercial reasons for getting this wrong. Jon has far less to gain in his report.

So, this is both a problem and an opportunity, but first, consider the problem side:
My wife loves Anti-static Fabric Softeners (Snuggles). I can't get her to stop using them and today I found "four (4) Snuggles" in the clothes basket. If you read the warning label, you notice that you should only use the "low heat setting" when using them and "avoid over dosing." Dosing? Why dosing? Because they are loaded with chemicals that should really never come in contact with your skin, ever. Unless you are trying to preserve yourself with formaldehyde.

From Wiki some of the chemicals are (quote):
"Contemporary fabric softeners tend to be based on quaternary ammonium salts with one or two long alkyl chains, a typical compound being dipalmitoylethyl hydroxyethylmonium methosulfate.[3] Other cationic compounds can be derived from imidazolium, substituted amine salts..."

Let's pick one of these "imidazolium," This, the smell good component, acts as both an acid and a base. It is synthesized from formaldehyde. Do you really care that much about static cling as to subject your skin to this chemical?

Skin care products and cosmetics are no better... and women, especially, actually wear them on a daily basis. I could go through any one of them and pick out a chemical, give you its basis, and show you how it causes Toxic Overload of your Liver and negatively affects your health in any number of ways. You can do the same within five minutes, using Google.

Think of transdermal migration as the "syringe injection" of a chemical. It goes straight into your liver and toxically overloads it, bypassing the body's various intermediate oral digestive steps. Liver overload means Toxic Stress, which leads directly to cardiovascular disease and cancer over time. All of this for eye makeup and clothes that do not cling. Sounds like a poor choice to me... agreed?

Finally, on the problem side: if you simple washed your arm pits really, really well with a wash cloth, chances are that you'd never need a deodorant... I discovered that years ago. As Jon says, some of us are worse than others. If necessary, do it twice a day, but avoid deodorants when possible. They are all aluminum injections and the arm pits, especially, are very susceptible to transdermal migration as Dr. Sircus points out.

On the plus side. face and body creams can become the most healthy delivery system for nutrients (beyond transdermal magnesium) ever devised and I intend to take advantage of that soon. So covering up "fine lines and wrinkles" will no longer be a required... you can literally take your nutrients transdermally and become younger visually... all in one process. We are making progress.

Kind Regards, Huuman

Toxins in Cosmetics and Shampoos
3/10/13
Hi Melly:

OK on the Coconut oil... now let's talk about the shampoo (and cosmetics):

This is about carefully reading labels and not buying commercial products that start off with ingredients like: Sodium Lauryl Sulfate <http://en.wikipedia.org/wiki/Sodium_laureth_sulfate> (SLS). Aluminum Lauryl Sulfate <http://en.wikipedia.org/wiki/Ammonium_lauryl_sulfate> (ALS) and contain Parabens <http://www.buzzle.com/articles/paraben-free-cosmetics.html> such as Buylparaben <http://en.wikipedia.org/wiki/Butylparaben> , Ethylparaben,Methylparaben <http://en.wikipedia.org/wiki/Methylparaben>

, Benzyl-parahydroxybenzoic acid, and Propyl-parahydroxybenzoic acid. These chemicals go by literally hundreds of pseudonyms. Few commercial shampoos made today, if any, do not contain these. It takes only a few minutes now on the web to research these products. If you do not recognize the names on your labels when you read them, look them up on Wiki, read the links... learn what they are.

Many people don't realize just how powerful (read poisonous) these ingredients are. The reason that the FDA turns a deaf ear. Supposedly, their rational is that a shampoo only contacts your scalp for a short period of time. Also, the amount is kept down as a percentage of the overall contents, but no one apparently controls or even publishes percentages. Repeat applications increase the absorption of these toxins. In fact, these shampoos actually increase hair loss. Read my book for more on this.

Commercial cosmetic products, in general, are unregulated and most contain questionable ingredients. Toxins are well absorbed by the skin, which is why transdermal magnesium is such a powerful health protocol. When applied to the skin, they go straight to the liver. Strong toxins, of course, greatly compromise liver function. Furthermore, these toxins add up, so avoid even small amounts at all cost. Finally, we get way too many toxins in things that we cannot avoid. Cancer rates are increasing for a reason. "It's the Liver Stupid."

Paraben Free <http://www.buzzle.com/articles/paraben-free-cosmetics.html> cosmetics are now available from most supplement suppliers.

Regards, HUUman
>Sometimes the best time to do this is before shampooing. Slather scalp and hair if you like, wrap with hot towel or plastic shower cap for 30 minutes at least. Shampoo. Repeat doing this and you will see good results after a while. Also ingest some vco. It will also lead to healthier skin. Start at 1/4 tsp then work it up to 3 TBSP per day. Never take vco on empty stomach - bad painful diarrhea will take place. Also don't take it after 3pm, too much energy won't put you to sleep.

Melly<

2/14/13

Bone Mass/ Raspberry Ketones

Hi Barb:

I find the results very, very questionable. First off, who sponsors this Science Daily site? Reading thru their stuff, I would say Big Pharma. Just a few things that pop-up that make me wonder:

1. These Raspberry Drops (Raspberry Ketones) are extracts of raspberries and have little or nothing to do with bone mass, from what I have read. How could they? It does not make any sense.

They say that there is a correlation between skinny rats and bone strength (if I am clear on this). That would be what you would expect with weight loss: (Snip)" "Circulating adiponectin levels are significantly lower in obese humans and rodent models than in lean controls. It is known that excess body weight and elevated body mass index are strongly correlated with high bone mineral density, and that weight loss is associated with loss of bone mineral density and increased risk of fractures. However, the mechanism for this relationship is unclear."

Here is the reasoning: Heavy people are stronger (rats in this case) and must have higher bone mass just to support their excess weight or else they can't even walk... that is a known fact. If they lose weight, their skeletal mass decreases and they lose some strength... duh! They simply don't need it and its all relative. This would not be the hormone related bone loss that women experience post-menopause and what older women should be concerned about. This is why you must do strength training and interval training as you lose weight... you'd like to keep both... And... lose weight.

2. The "scientists" that they reference on this site also tell us that (Snip): "Monitoring bone mineral density in postmenopausal women taking osteoporosis drugs (bisphosphonates) is unnecessary and potentially misleading, concludes a study published on the British Medical Journal website."

From all that I have read, bisphosphonates create nasty bones that seem dense, but are porous and break easily despite their density. It would appear that these reports all come from drug companies that are selling bisphosphonates and they are completely bias in their findings... plus they do not want you finding out just how bad their drugs are, "So please do not monitor them."

Certainly, I like their idea of taking women off of drugs for three years after they have been on them given what unbiased findings report... why not? There is nothing good about them and we have much better ways of increasing bone mass and stopping osteoporosis... magnesium supplements are one of them. They are cheap and effective.

There is a plot, but you are following the wrong trail here... It would appear.

I am not a huge Dr. Oz fan, but he has led you to a simple weight reduction modality that people on SAD diets are going to be disappointed with. Of course, the mainstream drug companies hate natural cures, so why not try & find problems with it?

Does this modality work? Probably not very well, but we know that there is no magical cure for weight loss. You simply a eat the Low Carb Diet with plenty of saturated fat and greens, add Whey, and MSM.... and everything just falls into place. Certainly, the Drops (or Raspberry Ketones) that Oz talked about are not likely to ever affect your bones either way.

Let's hand it to Oz for working, at least in a small way, toward natural cures. There are so few out there .

I actually am totally uninterested in weight loss, but I did buy a bottle from Swanson... for $3, it seemed like it would make a nice addition to my smoothie, on occasion and raspberry ketones are hard to come by otherwise.

How did they FDA approval? Good question. They didn't. This is a supplement and that is the rub. Drug companies pay millions to get approvals for crap like bisphosphonates that actually have negative effects and leave you with porous, dense "looking" bones and make millions advertising and selling them to people like you. Don't fall for it.

Regards, Huuman

>Very interesting. The plot thickens...
>Barbara

On Thu, Feb 14, 2013 at 7:54 AM, jadedhills@ntlworld.com> wrote:

>> Causes Osteoporosis
>> http://www.sciencedaily.com/releases/2011/11/111101171036.htm
> >Regards Peter

>>>Barbara Shen
>>>February 13, 2013
>>> *To:* Beck-blood-electrification@yahoogroups.com
>>>Subject: Raspberry Ultra Drops
>> >Do anyone of you believe in this drop?
>>> How they get FDA approval? How did they get Dr. Z to endorse this?
>>> http://www.fox5.news.newerfx.com/?s=lf

Arthritis and Fungus... Any connection?
2/15/13
HI Mike & Helen:

I am certainly no expert on mold (let's call it fungus... more general) and I am not really certain what you mean by it nor even where you are affected by it. Are we discussing a Lung Infection? Maybe, if you would better clarify your condition, we as a Group could get a better handle on what we are trying to cure, but here is a general response:

As to the Jim Humble MMS Protocol (and its not so new)... basically, the Chlorine Dioxide Protocol that Mike alludes to and as many here are familiar: Jim has been pushing it for years as a universal treatment. For certain things, it has worked for me as a short term internal treatment for problems like Candida/ Thrush. I would never suggest it long term and I would argue that it does not just kill the bad guys, it will kill all fungus and bacteria. In fact, Jim's argument that it does not kill them all makes absolutely no sense and is totally counter-intuitive.

Chlorine Dioxide would make a great spray as Mike alludes to kill mold in your house. Its safe than laundry bleach, but mold hates laundry bleach also. By the way, don't confuse MMS with MSM... they are unrelated. As an architect, I am called on occasionally to address mold issues in basements, etc. They generally are caused by moisture traps of one kind or another and a spray bottle of bleach water is a great first move. But as we know bleach is not so good to be around either, so chlorine dioxide is better as it dissipates quicker, as Jim Humble tells us. Either one will get it though. The key is to get a good ventilation path. By the way, I live eight miles from the beach, so it is a problem here too. Nevertheless, you want a strong immune system to combat such problems before they get a toe-hold.

Back on topic: Does MSM kill candida and other fungal varieties? I have no reason to believe that it would. Just as I would not suggest that eating a great Low Carb Diet or Whey would. However, I also don't get the relationship between "fungus" and osteoarthritis... assuming that is what we are discussing, really. I am thinking that someone is extending this fungus' ability way beyond its effects. I have never heard anyone attribute a fungus to any major common disease except cancer and I disagree that it is the "cause" here. Rather, the fungus thrives because you are so physically depleted that your body can no longer reject it and you can't Thrive.

I do agree that probably 25% of us do get fungus growths... probably a lot more, when you include skin infections and thrush. However one problem that mainstream medicine seems to have is separating cause and effect... especially with heart disease. This just sounds like another glaring example.

As far as getting rid of osteoarthritis, you are definitely on the right path. As I believe that osteo and cancer are not really even diseases, but are actually your body's reaction to toxins and poor nutrition.... SAD. I know, first-hand that HL MSM, per my protocol, will reduce osteo within about two months once you are on the full protocol. In fact it will generally repair sports injuries and injuries resulting in any trauma in that same period. It is also great after unsuccessful operations to get the body back inline with what it already knows it should be like. As a Quantum Entity, your body is one smart critter and it knows how it should look and feel. Your tiny little brain may argue, but it will lose, if given the debate, with the help of HL MSM.

But the really good news is that the AHP, per my book, gets your liver working like it should and quickly. That means it is a happy organism, period. Happy organisms are less prone to any kind of infection, fungal, viral, or bacterial... no more Toxic Overloads. More Glutathione, the Master Antioxidant that Duncan speaks of often, and a powerful Methylation Pathway with plenty of Organic Sulfur make the liver smile. This, IMHO, is the most healthy path that there is to a long and full life on earth. Nothing else comes even remotely close to it.

Recent Story: I have a friend, Billy, who was in an 80 mph car accident who dislocated both shoulders and messed up both knees some years ago. After one month on this AHP, he is well on his way to a painless, normal life. He still gets twinges, but he is back in the gym lifting and living a normal life. So its not just arthritis that it cures. He never believed it could happen... and without the AHP, it would not have.

I welcome all feedback from the Group on their discoveries... good and bad... following the AHP. We need to hear about them. Post them here

Regards, Huuman

From Michaelinde:
>I am not jim and don't have the knowledge base he has, but thought I would share this tidbit

with u. Some time back was doing some research of this new product coming to light called MMS. I don't use it but some of the claims are, mold is destroyed in dwelling places and some apparently use it . There is a yahoo group I believe.
Mike

>>Hi, Jim

Just finished your e-book you sent to me. Thanks. I have started taking MSM one Tb. spoon a day and will try increase gradually.

Does MSM can detox MOLD? Having been living bay area for 30 plus year and a moldy apt. for long years, I have all sort of health issues including deep soreness arthritis like bone pain/muscle pain all over. I was just been tested having very, very dangerous high level of MOLD from an naturopath doctor. He said 25% people have problem detox mold and I was one of them. Mold can cause all sorts of health problems. Overheard famous health author Andreas Moritz died from mold toxin last year. I think mold indeed is a big threat to our health. Besides changing the environment to move to less moldy place, I want to full force focusing on detoxing mold. I assume MSM is one of the good tool for that? Love to hear our group discussing this common mold problem issue.

Helen

Caution with DMSO
2 15/13
Hi Randy:

My thought is that I would not want any kind of soap in my blood... and that is what you are ultimately asking.

DMSO is potent stuff and it is dangerous, as I have warned many times, if you are not extremely careful with it. Think of it as a, "Transdermal Syringe." That is as accurate a description as you will get.

Yes, soap will emulsify oil and oil will not mix with water based solutions. My next question is: Would you inject hemp Oil into your blood? I would not. It may be great stuff orally, but directly in your blood? What are you thinking about Randy? Soap will also emulsify blood... and probably kill you if you get enough in there. Hemp oil's effect is totally unknown, but it could also kill you when it is injected.

Have I opened a can of worms here? By the way, don't try injecting hamburger into your blood either... just a subtle warning... and yes, you can eat free range meat without bad effects. That is not the point, though, is it?

I am repeating myself, but anything delivered through the skin, Transdermally, goes straight to the Liver. If you overload the Liver beyond a certain degree, you can expect major organ failure... the kidneys might actually shut down.

This recalls the time that I washed my hands in turpentine and my kidneys started shutting down. Turpentine is not acetone... we consider it pretty natural stuff (you know from pine trees) and I had washed my hands with acetone with no immediate effect previously, but I knew enough to get it off ASAP . The liver hates turpentine more. By the time I got to the sink to wash it off, I had nearly collapsed in a heap... and all of that was just through the skin on my hands.

Please be careful!

Regards, Huuman

Editing note: The fact is that without Surfactants, we would all die. However, DMSO is not a surfactant and never has anyone shown it to be poisonous from any perspective. Also, several group members have used it in their eyes as I have, It is safe, but always go slowly. This stuff is hot when 100% and will kill insects at full strength (4th edition)

DMSO

Hi Jim,

I've tried to mix DMSO and hemp oil, but they do not mix together. Same with DMSO and essential oils, such as melissa oil. I have been trying to think of some possible emulsifier for the hemp oil, that I don't also mind having carried into the body with the DMSO. I tried Dr. Bronners Castile soap. This does work, but not sure about having that absorbed heavily into the body. Any ideas?
Thanks,
Randy

Sun Feb 17, 2013 3:25 pm (PST) . Posted by: "Jim" huuman60
Hi Luigi, Randy & Mike:

No, I was not aware that Dr. Gregg had died, His study was passed to me via an email years ago along with a link and I pulled it up for my book. But obviously, he and I agree and he had a lot more credibility as a PhD. than I do as a reporter, so I used his work to verify my claims .

I have used it, in helping others, in delivering various nutrients and it has worked well in the past and I used it once in treating myself. So I know that he was correct in his basic ideas.

Dr. Gregg was a very intelligent man, obviously, but not very good at explaining himself. It took a lot of work just deciphering what he was getting at, when reading back through his papers, but I think that I finally got it down, LOL.

However, Randy may have encountered a flaw in his claims, here. It may not be so soluble in oils... at lease in certain oils and I have not yet tried it for that, but I do plan to soon. I can tell you for sure
that I'll have a better idea on that one, first-hand.

Good one on the K2, Mike... LOL.. I was all over that. Also, true on the Japanese being against GMO's... in fact, most intelligent people are, but they are not trying to make Cargill & Monsanto happy like the FDA is. Good source, it would appear.

Let's hope that the Japanese supply all of the natto and that it is not affected by the nuclear waste that Dunc reports on, though. Neither one is good.

Regards, Huuman

>Jim/Huuman,
>Thank you for defining an acronym after it is first used, as you have just done so well, a good practice. By the way, do you know that David Gregg passed away some time ago? David believed that DMSO was an excellent "transport"; for what you wanted you wanted to carry through the skin, from Aspirin to Cesium Chloride.
>Regards, Luigi

>>Hi Jim, yes, I'm serious. Not sure why, but they don't mix. I have 99% pure DMSO from a
>>source in Tulsa, OK. I dilute it to 70% and 30% for different. Straight, it hard on skin. But
>>you have me thinking. Randy

>>> I haven't figured out how to buy Natto from anywhere except the Japanese shop... and the Japanese are notoriously anti-GMO. Doesn't need to... only needs to break down the products of those
genes. There's a good chance that fermentation does help make even GMO foods healthier.
>>>Mike

Hi Floyd:

There you go. Given that these natural remedies reinforce each other (unlike mainstream drugs). I would do both Peter's and the EMU oil solutions. MSM cream is just an effective delivery system.

You might even throw in the herbal mix in the cream. So here is what you'd get:

1 tablespoon Dried Fresh Sage
1 tablespoon Dried Fresh Oregano
1 tablespoon Dried Fresh Thyme
2oz MSM Cream
2oz Emu oil

Let me warn you about cutting corners on the herbals... don't. Buy them fresh from reputable suppliers who guarantee that they are from organic suppliers. Then follow Peter's advice on preparation. To start with, don't cut corners.

If this does not work, I have some other ideas, but try this first. Its still simple.

By the way, these herbals are going to look pretty bad on your arm, LOL. So you might want to start out using them separately and see how that works, first. So go through half a jar of the EMU and the other separately. If that does not seem to be effective, combine them for the 2nd half. If that does not work, get back to me.

I will be doing some serious testing on skin creams over the next few weeks, but not for this. I, most likely, will make some serious discoveries... they always pop up somehow. If they are too cool to reveal, we can talk privately. Heck, we might just come up with two great cream solutions out of this, one for rashes and the other to reduce aging.

You may begin to feel like a test dummy, but you may make a discovery that you can use forever on many things like Peter has done. I have never really used the Fulvic Minerals and Clays, per say, but this would certainly be one place that they should do some good.

Are you with me big guy?

Regards, HUUman

Supplements
2/19/13
Hi Floyd:

Yes, you are right about the K2.

The technical information in my book can be deep and unless I review it, I don't get it either. However, the protocols are very, very simple to follow.

In my opinion, generally, there are far more essential nutrients that you must supplement than you mention here if you plan on staying well for long. By the way, in no place do I mention vit C, and I have my reasons. I stopped taking it maybe 8-10 years ago. However, I eat a lot of fruit including whole lemons and limes (both very low in fructose) and I am considering adding liposomal C at some point.

My own regimen probably includes some 50+ supplements/day including, yes K2, but also B-6, B-12, Magnesium in various forms, Colloidal Trace minerals, MegaHydrate, 30 different enzymes, a lot of herbals (as Duncan as I have discussed here), Quinogel (liposomal CO Q10), etc. and a very careful Low Carb Diet, with lots more Whey than Duncan advises, and plenty of MSM. It varies according to what is going on and what I research. It works really, really well for me, because I have absolutely no health issues and I just turned 71. I play tennis three to five times a week and life is good. I will be living well till I die... that is my plan and sickness is not a part of it.

It is more about discipline than "learning" Floyd. Look at your issues and address them separately and together. That is the point of my book. Do your research and build your own health program. My book can serve as a skeleton for that approach, but it gives you three very valuable things to start with: MSM, Duncan's Whey (along with Duncan's suggestions), and The Low Carbon Diet. It also describes why they all work in unison and separately... take it from there.

If you have specific issues, I am here to help... as is this wise (note I did not say learned) group. There are plenty of really learned people who don't have a clue... and often they are the problem. I also suggest that you join the Coconut Oil Group, because it is equally wise.

Kind Regards, HUUman

>I have been reading all that have been put on line here and have a problem with my arm. There is an catch that I just can`t get a handle on. I belong to this group and have downloaded the ebook that you sent out to all. The problem that I have is that I am not much to these kinds of instructions. I don`t have the learning that most of you have. I have read the ebook and can`t

get hook on it to understand what is being said. I have been taking MSM and vitamin C as some instructions I have been given. Now I see that you say to take K2 with it. Am I right about that? I am just having a problem putting it all together. Thank you Floyd Mackey<

Diagnosis
2/20/13
Hi Floyd:

Go to a doctor and get a diagnosis for the arm so that we can comment on your problem. If no one can give you a diagnosis, maybe you could describe the actual condition in detail and someone here could relate to it.

As to calcium build-ups. I am sure that there are plenty of opinions here as to what causes free blood calcium and related build-ups on joints, but I am equally sure that the AHP itself, as outlined in my book will help you... it may fix your arm too, but that one remains a mystery. Certainly, if you have a parasite or something lurking there, it will have no effect.

After you have been on the MSM protocol for a month after the start-up, get back with me and let me know how you feel. Make sure that you pump as much Magnesium into your body as you can via oral chelates and transdermal oils. Other than that, you are good to go. I am expecting a really good report. Within two months, it should be about over, forever. That is what I have seen and heard so far. No knives are required. Thanks for asking...

Kind Regards, HUUman

>I don`t have control of my legs that started about 10 to 12 years ago. I was to see a specialist in fact 2 of them they both told me the same thing. There is calcium built up where the nerve endings go through the pelvis and are chocking the nerves the feed the muscle in the leg and I am losing control. If I could find away to get rid of the blockage I would get that control back. They to put me under and go in with knife and I would not let that happen.
Floyd<

Fracking
2/21/13

But we are getting rich!!!! So who cares, right?http://www.organicconsumers.org/articles/article_27073.cfm

This article is just giving you small a part of the larger picture. It is far more bleak, Its about

CAFOs and Factory Farming, making the U.S. the worldwide energy kings again, and sacrificing our future for today's gain. It is politically correct by today's standards and it speaks directly about how we have come to live as a nation.

This is, without a doubt, the most harmful long-term environmental practice ever... even worse than underground nuclear testing. As these first complaints come in, we begin to see just how precious clean aquifers are. No amount of above surface nutrient waste can come close to what this practice does and there are no known ways of amelioration... or any in sight. Unlike the poor farming practices that we have slowly come to depend on in the last sixty years, no one can just stop fracking and go back to sound judgment. It will take thousands of years for nature to clear this up, maybe, no one even has a clue how long. What we do know is that polluted aquifers make lands within miles of a site and sometime within ten mile segments unlivable for everyone just as these farmers report.

However, the Washington environmentalists generally turn a deaf ear to it and will continue to do so. Why? It is making the U.S. temporarily rich. It allows us to spend money that we would not otherwise have. What a deal! Politicians no longer have to make the hard decisions that they hate to make and they can actually fulfill their stupid promises that they thought were lies even as they made them. Cake and Eat it too.... you bet!

With fracking, we can once again become the energy leader of the world which means that we can virtually do whatever we want internationally. We no longer have to depend on the Middle East for oil and we can ignore those idiots that we have angered in the past. We do not have to drill in Alaska. We can make the "environmentalists" really happy and both sides can coexist politically. On the surface (literally) everything looks good. That all works till you drill a well for your farm or a raccoon drinks water from a spring and slowly dies just as the goats did on the farm down the road. The politicians must love seeing the 35,000 people march on Wash. D.C.. Hey, Nice diversion.

Strange what politically being Green means compared to practicing what works responsibly. With fracking, our energy resources can multiply geometrically and it is occurring almost over night. Dry oil wells in Texas are now beginning to produce just like in the old days and hundreds of miles of tar sands become liquid gold. Wall street can start counting their money and everyone knows it. Hey, we are rich!

Long term, it means, just as these organic farmers report, we can no longer even raise animals outside of factory farms. It takes a lot more land to raise animals organically and fracking will quickly eat it up. Our very future as a nation is at stake, but no one is blowing any whistles. Certainly, any chance that we had of becoming a healthy nation again becomes more slim by the day. We just will not have the land area required to support responsible farming practices.

The press is giving this a deaf ear and the politicians on both sides are happy. This is how we have come to live and... fracking is speeding up the process by leaps and bounds.

Skin Rash
2/22/13

Hi Floyd:

There you go. Given that these natural remedies reinforce each other (unlike mainstream drugs). I would do both Peter's and the EMU oil solutions. MSM cream is just an effective delivery system.

You might even throw in the herbal mix in the cream. So here is what you'd get:
1 tablespoon Dried Fresh Sage
1 tablespoon Dried Fresh Oregano
1 tablespoon Dried Fresh Thyme
2oz MSM Cream
2oz Emu oil

Let me warn you about cutting corners on the herbals... don't. Buy them fresh from reputable suppliers who guarantee that they are from organic suppliers. Then follow Peter's advice on preparation. So, right off the bat, don't cut corners.

If this does not work, I have some other ideas, but try this first. Its still simple.

By the way, these herbals are going to look pretty bad on your arm, LOL. So you might want to start out using them separately and see how that works, first. So go through half a jar of the EMU and the other separately. If that does not seem to be effective, combine them for the 2nd half. If that does not work, get back to me.

I will be doing some serious testing on skin creams over the next few weeks, but not for this. I, most likely, will make some serious discoveries... they always pop up somehow. If they are too cool to reveal, we can talk privately. Heck, we might just come up with two great cream solutions out of this, one for rashes and the other to reduce aging.

You may begin to feel like a test dummy, but you may make a discovery that you can use forever on many things like Peter has done. I have never really used the Fulvic Minerals and Clays, per say, but this would certainly be one place that they should do some good.

Regards, HUUman

Hi Jim,

I had a thought, I use Hemic/Fulvic acid as a general pick me up, I used to take it all the time, too expensive for that, but very strong anti Bacterial/viral/ulcerative. Try drinking some on an empty stomach for about a week and a half, see if that will fortify your system, Also for years I have been curing Skin rashes with the following as a wash:

1 tablespoon Dried Fresh Sage
1 tablespoon Dried Fresh Oregano
1 tablespoon Dried Fresh Thyme

Steep in ½ pint of boiling water till it cools to just warm, strain and decant, keeps for 3 days. Apply to the skin area 3 times a day. All three herbs are very strong in their anti Bacterial action and they inhibit the salmonella infection.

Kind Regards Peter

keys Speak
Natural Solutions <newsletter@naturalsolutionsmag.com> (link to site)

Hi Everyone:

I was just reading above on-line Article that was advertising Mushrooms of various varieties and touting their health benefits... which are truly great, especially considering the fact that the are generally white or brown as opposed to what we have come to think of colors in our food as the super phytonutrients that we should be looking for in our veggies as they point out.

Where they stumble in their article is when they begin discussing heart health and cholesterol reduction. Everyone, even alternative heath people, pander to the mainstream when they are pushing their products, so they still address their products from the Dr. Ancel Keys viewpoint or "Keys Speak" even though we know this guy was truly a quack in the true meaning (Not in the sense that "quackwatch" uses... That is, out of the mainstream's fixed-brain, fifty year old viewpoint on how things work).

So what they begin with is that their mushrooms "Lower Cholesterol." Now generally, what they are actually saying is that they increase HDL and decease LDL carriers. Thee carriers are not really cholesterol at all... But higher HDL is actually good. The liver is the primary mechanism in determining this ratio and when the liver is producing lots of the HDL, it is telling you that it is happy with no inflammation, toxicity, or fat. This is something in "Keys

Speak" that you have to get your head around when you are determining what you really want in your diet. In fact, no one following mainstream recommendations has a clue as to what a good diet is, so they don't know where to start.

So when a biased add relates back to studies... you research the studies, then you discover that their product apparently does reduce cancer and heart disease. What is that actually telling you? The chances are that the translation is that you have discovered a product that is reducing Liver Stress. It is that simple. However, to do that, it must be contributing to methylation, glutathione production, increasing your sulfur stores, helping your mineral balance, adding antioxidants and/or enzymes, or reducing inflammation... to reduce what it has to be doing to the nutrient level. (Per the Title of this book,) "It's the Liver Stupid," it generally starts there.

Mushrooms are foods that you want in your diet and, translation, they affect your liver positively. So, yes, include them. What you absolutely do not want is something that actually "lower your cholesterol." Cholesterol is essential for every organ beyond the liver, where it is Produced (and all other organs are beyond the liver) to function properly... Especially your brain. Low cholesterol diets are a big part of our population's ill health and most diseases that are increasing in numbers like Diabetes, Heart Disease, Cancer, Alzheimer, MS, etc. are at least helped along by a low, wrong-fat, diet. Of course, there are other factors. However, when the mainstream talks about "good fats" they generally still mean wrong-fat, or vegetable oils (but now often include olive oil). The point is that if your translation of "Keys Speak" is incorrect, it will be disastrous, long term.

We eat pork from pigs that have been bred to produce a very low percentage of fat compared the free range pigs of sixty years ago. Our chickens are bred and fed lean, primarily corn and soy bean GMO grain diets, so are our cattle. Their meat is all relatively flavorless by comparison. Why do we like the taste of fat and the free range nutrients that these animals normally got in their diets? Because it is good for you. Your body knows even if your mind forgets. Our bodies Thrive on good nutrients and give you sound feedback. Learn to listen.

As Dr. Seneff points out in her interview, all meat eating animals go for fat first, then organ meat, when they kill their prey. Cattle and chickens love to graze. Why? Because this is where the nutrients that they (and we)Thrive on are located. We must have them in our diet which is why we must chose our foods wisely and supplement our diets correctly. Coconut oil is one good example, but there are plenty of other examples, such as mushrooms.

Even the highly touted "Grass Fed Beef" that the alternative community speaks of is not up to the nutrient capacity of a wild deer or buffalo, because all of our farm animals have been bred to meet the Keys standard as much as possible for sixty years. To get back to where we were nutritionally three hundred years ago is probably an impossibility at this point, but at least we have to learn "Keys Speak" and learn how to translate it into as healthy a Low Carb Diet as possible.

In Health, HUUman

MSM Bitter Taste
2/24/13
Hi Theta:

Technically, you are absolutely correct, Theta. There really is no bitter taste associated with MSM and here is why:

Theoretically, methylation that occurs when you first put it in your mouth sets off a chain of events. These overload the mitochondria and they, at first, cause what some call "Bitterness." Bitterness, Salt, Umami (savory sensation), Sourness, Sweetness... these are tastes. MSM actually has none of these. But by overwhelming the mitochondria until they are put to "sleep" by the methylation, some just feel "horribly uncomfortable" if they swallow and this discomfort can actually cause some to vomit uncontrollably. That is an extreme case, but it can happen. For others, at the other extreme, there just is no problem from day one. Most are somewhere in between.

The key is to keep the MSM and water in your mouth till the mitochondria are basically intoxicated with this methyl component. A similar thing happens with any strong alcoholic beverage. The first shot is makes you wince. After that you can drink down a shot of pure 100 proof scotch and nothing happens taste-wise. In both cases, you have put your taste buds to sleep. Taste buds are specialized receptor cells that allow information to enter the mitochondria and finally the brain.

Some of us have many more than others. I am a supertaster and have always had problems with bitter things, but MSM was never a big problem for me. Others may taste very little and MSM drives them crazy unless they follow my advice. In fact, some people really cannot taste much of anything, yet they may still have a problem if they just try drinking down MSM and water without the swishing. I have a friend who can only taste hot peppers. He has not gotten on the protocol yet, but he would be a nice test case.

I had an injury some years back when I hit a coral reef. The boat almost went dead in the water when it hit. The bridge of my nose struck the wheel of the yacht I was sailing and that interrupted the channels from them. I lost both smell and taste for six months. You grow new taste buds every ten days or so, but this nerve channel is key to keeping them all working, so it

took six months to fix itself. Check out the basics of taste at:
http://www.plosbiology.org/article/info:doi/10.1371/journal.pbio.0020064

My theory is that the mitochondria are the key to all of this and, hence, the methyl groups that MSM contains works as it does. Most taste scientists (yes, Olfactology is a field), have not gotten this far, but a few have. Given this theory, one of the lesser benefits of the Anti-Aging and Health Program (that I describe herein) is that it will reactivate the mitochondria and allow old people who have lost their taste sensitivity to get their taste back, but the other benefits are obviously far more important. Nevertheless, having gone through six months of being tasteless, I now fully appreciate being able to taste things

Kind Regards, HUUman

>I have heard that MSM tastes bitter many times. It is tasteless to me. Anybody else find it tasteless?
I can put 5 grams in a little bit of water comfortably.

Theta<

Alzheimers
2/ 27/ 13
Hi bG:

Interesting and well written piece... this is the only one part that I find suspect and you probably already know why:

Dr. Joel Wallach claimed to have induced Alzheimers in animals almost twenty years ago by depriving them of saturated fats in their diets. Since brains are nearly all cholesterol and our SAD diets are nearly devoid of them today, it all adds up, but it has never been substantiated otherwise.

By using coconut oil in diets and eating animal meat that have not been purposely bred in an effort to reduce their body fat to lean meat ratios, and eating foods, in general, with natural fat ratios, Alzheimers's should become rare gain and basically eliminated from our list of "modern" diseases. Wallach calls it a, "doctor induced disease." If he is correct, and I think that he is, the herpes association that you reference has nothing to do with it.

The disease, then, Alzheimers, although named after the doctor who first "discovered" it, should be renamed the "Keys Effect." The Keys Effect is not really a disease at all, its an nutritional abnormality. However, most "Dis-eases" are. In this case, though, one man Dr.

Ancel Keys, one of the really dark characters in our health history that the mainstream still loves, was the culprit who has bought it on.

By the way(as I reported herein), Dr. Keys also invented K Rations... in his brilliance. For those who love sugar, it was the mainstay of American Soldiers on the battlefront during WWII when things were really bad. It is comprised of hard candy, hard biscuits, dry sausage, and chocolate. Knowing this, it's difficult to fathom how we actually won that war. and Furthermore, is a good thing that it did not last as long as the Afghan War or the soldiers would be dying left and right of malnutrition. However, when you get right down to it, our entire population generally dies of Cancer and Heart Disease which are forms of malnutrition. I also noticed that many retired Airmen died of heart attacks immediately on retiring. That means at about age 40 to 45 on average, Given the "Chow Hall" diet that we were fed, it all makes sense now.

Regards, HUUman

>The Lyme's bacteria may not be able to stand the metallic silver particulate in its environment, so leaves the brain and nervous system once they become seeded with silver. It could be a near-permanent protection against not only Lyme's but also herpes, which has been associated causatively with Alzheimer<

Selenium and Key Mineral Nutrients.
2/27/13
Hi All:

Adding to what Duncan has said below:

Selenium has been shown in cancer research to be a strong cancer inhibitor (and often)... and for good reason. That is, there is a strong correlation between levels of cancer incidence and lower levels of selenium in the blood because of the pathway that Duncan mentions.

Furthermore, like most of the minerals, Chromium, Magnesium, Silicon, Potassium, Sulfur, Iodine, Boron, Zinc and to a lesser degree Vanadium, modern farming practices have taken Selenium and Trace Minerals from our diets.

The key thing here is that plants do not manufacture minerals. If they are not there naturally, you just do not get them ever... and you get sick. All disease can generally be traced to a lack of one or more of these. Add to this that modern practices require long shelf lives and you multiply the problem.

The three most important of these are Magnesium, Sulfur and Selenium: Magnesium for Bones and Heart and Selenium as a hedge against Cancer.

Sulfur is the most volatile of all of these and it is probably the most important element for health that we are all lacking in our diets, but who is counting? You can't survive long when any are missing. Any food that has been processed in just about any way, is devoid of sulfur. Even refrigeration, or just sitting around, destroys sulfur content. For that reason, no one, likely, gets as much as they could possibly use.

Sulfur is the "key element" in the electrical transmission between cells. Does that tell you anything? There is more... a lot more, but the idea here is that you must first make sure that you are getting all of these minerals back into your diet that modern practices not longer accommodate.
Foods designed to intentionally sit on a shelf, be refrigerated or frozen, or are otherwise processed, can never deliver sulfur. So you must, somehow, figure out how to get it back into your body.
MSM, which also provides an elegant system for methylation (look it up in my book) and organic sulfur, and Whey are two superior delivery systems that Duncan and I speak of. But there are other, still important, adjuncts that also help. Selenium is one of them.

Finally, we must keep this glutathione conjugation pathway open that Duncan speaks of here or else the liver becomes overloaded and the entire organism (you) dies... as this death would be from what modern medicine would term an unrelated disease like CVD or Cancer. Fundamentally, this may not be easy to understand, but it's simple to take care of when you know what you are dealing with.

Kind Regards, Huuman
Mon Feb 11, 2013 1:05 am (PST) . Posted by:
"Duncan Crow" duncancrow
>Selenium is primarily used in glutathione production by all the cells >depending on how much they are exposed to toxin load, and a bit is used >by the thyroid. Can't be well without it; dose is usually 200-400 mcg, >depending on arsenic loading from the drinking water.

>all good,

>Duncan<

MSM Problem (in the order sent & received) :
Mar 1, 2013
Mike wrote:

Hi, I've tried MSM in the past but had to stop because all it does is make me sore. It feels like straining a muscle and I pull up lame. Obviously you can't go through life limping along so I stopped taking MSM. Have you heard of this problem before? Do you have any suggestion? I have high copper levels... and this may be some of the reason my body responds strangely to some supplements.
Posted to groups and sent Mike:
I certainly think that you are on the right path here. The antagonizes giving you a reaction would be one warning and your sensitivity to MSM is the other.

Next, as you say, You have to find your sensitivity threshold and stay within it till you get this licked with all of these, but they have to be key to beating it.

Good that you got a diagnosis on the Wilson's Disease... of course, I would not know the term, but there is a disease associated with each mineral in the same way... and that was my first assumption. How did your doctor determine that you don't have it? Just curious.

Not that this will make you feel any better, but probably five hundred are on the *HL* MSM and you are the first that has such a problem. However, it only makes sense that this problem is the key to your beating this. You just have to get it right and stay with it.

You might want to try some flushes and cleanses to see how/ what they do for this. Jon Barron has a lot on that at his site about this. Again, take it slowly. You might start with a very light liver flush and see what that does.
HUUman

Continued Conversation from above email question:
Mar 2, 2013
H> Sounds like a strange detox reaction.
Mike> That's been my take on it for a long time.
H> Realize that a healthy body makes MSM naturally. If you are detoxing, and that is about the only explanation, you have to go very slowly and get past it.
Mike> OK
H> How did you get the excess copper?
Mike> No idea.
H> Can you avoid it?
Mike> I'm pretty sure it's not coming in via the diet... as nobody else in my family has high copper.
So either I've picked it up somewhere or I'm absorbing it particularly well.
It's officially not Wilsons disease (the genetic disposition to retain copper) according to the doctor (I got a second opinion just to be sure).

H> There has to be a solution.
Mike> VitC, manganese, molybdenum, zinc all antagonize copper.
Manganese, zinc and molybdenum all make me feel unwell if I take too much of them (doses that don't make normal people unwell). It appears that MSM is another thing that antagonizes copper... that I hadn't worked out. I've been chelating copper for the past 7 years. Only VitC (only the soluble form works) doesn't make me feel unwell... so I've favored that... but do try to make sure I keep manganese, molybdenum and zinc levels up... but below the threshold that they make me feel unwell.
It feels like getting angina. In fact it feels like the minerals go to war. I'll try adding MSM on a regular basis as part of my chelation strategy.
H> Regards, Jim

Continued Conversation from above email question:
3/3/13
>Hi Jim:

Actually, this can be a quite typical response to MSM stripping acidic waste from the body. You can actually chart the effect by measuring saliva pH. As soon as the pH drops down below the 6.25 level, fatigue and varying amounts of muscle pain start to occur. For those with conditions like CFS, it's not soreness that is experienced, but pure agony.

In fact, for this reason, I sometimes use MSM as a diagnostic tool. It doesn't matter how well a person thinks they are or looks; an extremely healthy individual will barely react to taking things like MSM.

To help counter the effects, tissue hydration and liver support needs to be a priority. You can't get proper facial hydration without proper
mineral balance, and you can't properly balance minerals without proper soft tissue pH.

Someone who reacts very severely to low amounts of MSM (1 gram) should consider a really thorough daily cleansing program. For example, 1 tsp. of edible clay twice daily, and the juice of about four lemons a day (in high quality water).

...cancer is right around the corner for many people who react severely. There are too many environmental pollutants out there post 2000 to think that eating healthy will prevent cancer. The body's detox pathways must be mastered, as well as the body's metabolic state.

~Jason<

Continued Conversation from above email question:

3/3/13
Hi Jason:

Interestingly enough, the late Dr. David Gregg suggested using DMSO as a diagnostic tool (Ch27). His method was that since DMSO imparts a strong garlic breath on people with serious health problems: The smell of garlic breath gets stronger relative to their problem, compared to a well person and that can easily be measured. So the stronger the garlic smell, the worse the problem. MSM does not do this, but your point is well taken and it would be a good tool, though maybe more subjective, than an actual breath analysis as Dr. Gregg suggests. But why not track both if you are using DMSO? However, I normally favor MSM, just because it is safer, and your pH analysis sounded like a good tool the first time you posted it here.

I immediately sensed that Mike, was in bad shape as you suggest, from his description. I also realized that his mineral balance was way off as he virtually told us and, finally, that he has to get it straightened out ASAP.

I'll send him your take on it, Jason. Also, as far as hydration goes, its hard to beat what Dr. Patrick Flanagan has contributed in his work for that, but clays can be cheaper if they are done well as we have discussed previously. I would actually do both, given the gravity of the situation (can't call it an illness yet, maybe, but as you say, it is a step away from a very serious one).

I can also see where he is apt to come down with cancer (or something else) soon if he does not get past this problem. Mike sounds like a pretty hip guy who has tried to do his homework, but has just not come across the correct answers to deal with his problem(s) till now. I am thinking that he will be all over this when he reads it. This has been an interesting lesson for me.

I know that I have said this before, Jason... but nice having you around. And thanks for stepping up. Your input is absolutely invaluable. It is rather amazing that Mike's doctor did not "get it," but we know how that goes. If I were him, I'd be looking for a new one. With input like this, maybe we can begin to force the mainstream to revamp their priorities toward wellness. As we say in tennis... "Take the ball early."

Regards, HUUman

Joint Health is not the Basic Issue
3/3/13
HI bG:

Glucosamine, Knox Gelatin, Chicken Cartridge and Egg Shell Membrane are all collagen sources and if you could get enough (Hyaluronic Acid is another one of similar value), they all would at least begin to solve your problem. However, you can't and they will not do the job. Chondroitin is a waste of time, period, because the molecules are just too large to do anything. Dr. Wallach missed the boat on this one. He was only partly correct. None of these supply enough organic sulfur efficiently enough and none aid in methylation, but they are good adjuncts. I also recommend MicroHydrin, but this will not do the trick on its own either. There will be synergistic effects.

I am using the egg shell membrane as a collagen source (among others) in my DMSO delivery system, body/ face cream that I am testing, but this is a transdermal system and it is oriented toward the skin and is a supplement. They will help build the skin thickness back and thus take away the fine lines and wrinkles while in the process of delivering nutrients elsewhere.

You have a much more deeply oriented problem and I address it in my book. Joint issues always start at the Liver. "It's the Liver Stupid." To get the liver back you need a many pronged attack. You are not new to this, so I am sure that you are hitting some of the issues and you can still get around, so good, but you are obviously missing some others or your joints would not be telling you that they are unhappy, otherwise. Think of the joints as your "distant early warning" system, since they are last in the food chain and first to be hit, as the liver struggles to meet its quota and keep you otherwise well.

You are probably already doing most of what my book suggests but obviously are not on the HL MSM protocol (or you are taking creative short cuts as some others have been). It is like being two quarts low, but with an oil leak and screaming joints are valve clatter in a sharp turn. Certainly, there is no reason to be putting up with screaming joints, but they are telling you that other, much more important problems are coming unless you pacify them... the engine will eventually blow.

Kind Regards, HUUman

Joint Damage
Fri Mar 1, 2013 11:30 am (PST) .
Posted by: Baby Grand
>I'm going to try this for some hip and knee cartilage damage. So far I'm functional, and have just started using egg membrane capsules. I figure why not use the chicken and the egg, just sounds so complete.

I've read good things about these 2, and thought to give them a try. Has anyone else tried these 2 things? The glucosamine, chondroitin supplements I took didn't seem to do much. The Knox Gelatin helped a little more.

Since taking the egg membrane recently, things did seem to improve, but it's very soon, like 3 weeks, so not sure anything real is happening.

bG<

Paul wrote:
DEW. My dashboard is full of them. Very scary. Wish I could get one light off. But instead more keep turning on. Ugh. Hope to turn the corner soon. Thus is a very smart group thankfully
Sent from my iPhone

Hydration
3/3/13
>Jim,
Does MSM thin the blood? I notice sometimes my blood is nice and pink and thin. But today, it was thick and dark. Thanks and just to let you know. I am reading your ebook slowly, to make sure I remember.
Melly<
Hi Melly:

I believe that what we are really discussing here is hydration. The reason is: When I get Migraines (seldom anymore), it's because I have thick blood. My theory is that a migraine is the body trying to save itself by dilating the blood vessels and arteries and sending the available blood to the brain which is starved by poor blood flow. The first hint of a migraine is visual, followed by the left arm going numb. The progression strongly favors this theory as the problem. My Chiropractor friend, Dr. Hobbs, first suggested it years ago.

The next hint is that Mega Hydrate works by hydrating the blood and it fixes migraines. So my favorite answer to hydration is Mega Hydrate. Mega Hydrate goes into virtually everything that I drink... at maybe ½ capsule. I usually can buy it at around $25 for 60 capsules, so it is expensive.
The Crystal Energy (liquid) that Flanagan makes is far cheaper to use, but not as effective in my experience.

Also, as Jason suggested in an earlier post today in the Beck-Blood-Electrification Group, edible clays help with this. My suggestion is do both and do them regularly.

Regards, HUUman

Mixing MSM with Water Negates the Methylation
3/4/13
Hi Ivo:

I don't know anything about you, but if you are over about 25, this cancer tutor protocol that you have found is almost totally worthless for you. Also, your following this protocol had absolutely nothing to do with your cyst progression, IMHO. This mixture would be very close to a zero on the scale of one to ten for effect. The fact is that your body makes far more MSM naturally than this protocol delivers in just keeping you alive. My HL MSM Protocol delivers many more times the effect of this, once you are past the "start up." Keep that in mind when you get serious about it. The water dissolved delivery, where many make their first mistake, negates "methylation" almost completely... and we could go on from there.

I am sorry that you are having the pain, but you just can't blame this innocuous treatment on your progression... actually either way. Had the cyst disappeared, it would also not be because you were taking this protocol. You would simply have needed to look elsewhere for the reason.

MSM has no "bad" side effects, in itself, but has many good ones. Taking much larger quantities of it than you have taken, too fast, could cause toxic release (assuming you have stored toxins), but this scant amount that you were taking could never affect anything unless you were on the verge of death.

You did not tell us how much D3, but I am assuming trace amounts there too. However, D3 has never been shown to cause these problems either. It took over 100,000 iu/day over a month's time to cause the members of the asylum to become ill in the famously reported experiment done in the '30's and they all recovered as soon as they were taken off of that crazy, dictated protocol. If you read the Vit D Newsletter, it have been proven to have anticancer effects, but you have not been dealing with cancer by what you are saying.

I am sending you my book. Read it carefully and you'll have a much better understanding of what you are doing here. Then, get back with me here. We can all learn from what you are doing. Benign cysts are very difficult to treat wherever they are because the body just does not recognize them easily and there lies the rub. In fact, cancer would likely be more simple to deal with using the tools that we have. I do wish you the best.

Kind Regards, HUUman

>I used the protocol MSM/Vitamin D3 based on cancer tutor. Then when I made the water msm (mixture of ½ gallon of purified water
and 5 Tbsp of MSM) and I began to give just 1tbsp /day the cyst become hard and painful. I tried several times in one month after a time of rest to come back the cyst and no pain on her. My opinion is that the msm forced to retract the cyst and the pain. I don't know. I Never more tried to use it again. Then I need to understand the side effects that can occur and how I can jump these side effects or reduce them.
Thanks,
Best regards, Ivo<

DMSO is not an Irritant
3/4/13
Hi Randy and Luigi:

Randy, that was my point and we agree. As Jason and others have said here, DMSO can be dangerous when not properly handled, so people must be careful when using it. MSM is far more forgiving. I generally would never recommend using it without following a list of precautions and I like the idea of 30% reduction with clean water. However, you can purchase it that way in the first place. This reduction, though, clearly does not make it inherently safer if you do not take the very same precautions (I mention this repeatedly herein).
What I don't like is saying that DMSO "burns" things. It simply can't. Some have reacted to it with a rash (irritated if you will), but my guess would be that they have not been completely sterile in using it and they simply have driven nasty stuff into their skin... and, no question, it will do that. That is what DMSO is very good at doing. Aloe Vera would indeed sooth their irritated skin in any case,

We have to be careful in our language so as not to discourage people from using DMSO and MSM in intelligent applications. This oxygen transport pair, DMSO & MSM, are both key to some outstanding health breakthroughs that could possibly be shut down or suppressed by indiscriminate terminology. So far, the mainstream (and FDA) has done its best to suppress the use of these and for years now, because they cannot make money with them. However, both are better than any "drug" that big pharma has and in a myriad of ways. This will come out as time goes on... and people begin using them correctly.

Basically, DMSO is just a far more absorbable form of MSM in that it has not yet oxidized. Theoretically, when either one has entered your body, they transfer back and forth between forms as they are utilized as Dr. Gregg describes. As he says, they are biologically just one thing (if he is correct). MSM is just easier for non-professionals to control and neither, in themselves, can possibly harm you. They are not "drugs" and neither one reacts with drugs... the theory is that they make you younger by "methylation" and "oxygen transfer." Thus act as both oxygenator and antioxidants as they transfer between forms within the body... which, as it were, nothing else in the world does this that anyone is aware of, so far. By the way, without these processes going on at the mitochondrial level, you would simply die. They are key to life's function at the cellular level. It must occur whether or not you supplement with them.

Kind Regards, HUUman

Heat Treatment for Cancer
3/7/13
Jim, with regard to the repetitious cancer claims made by the marketer for bio-mats, the researcher quoted says they are never used to treat cancer by themselves but to augment cancer chemotherapy. So, a marketer's mention of bio-mats in this regard is essentially fraud A hot bath increases core temp by the 2 degrees required for heat shock therapy in under 15 minutes, and this can be checked with a thermometer. This is what Paracelsus used "to cure any disease."
>
> all good, Duncan<
Hi Duncan:
> I have followed your arguments and your have my ear my friend. I am not buying Lymover's claims. Actually, my Father was looking into this method of curing cancer in the late '60's with great hopes... and finally gave up on it, basically, it for the reasons that you argue. He was a man well ahead of his times, but he actually bought into Dr. Key's BS on cholesterol... and that killed him at the young age of 67... which I now consider middle aged.
>
> But then, why worry about curing cancer when we can simply avoid it with MSM, Whey, and Low Carb Diet? We apparently have this licked. No one should ever get cancer in the first place... same with heart disease. I am becoming more certain that these are just peripheral diseases that should never occur in a population that follows their good senses. Read on for why I make this seemingly ridiculous claim.
>
> I hope that I am not offending you Lymover, but I do believe that Duncan is correct and I do have a fairly deep background in this particular healing mode.
>
> On a related topic, I just got a very impressive early result from a woman named Mona, whom I have never met, yesterday that she and I are very excited about... are you ready for this?
>
> Just one year ago, Mona read about MSM here and went on what I now call my HL MSM Protocol. Now, please realize that I was working almost totally on the empirical level then, but was a big advocate. At that time, Mona was 53. She had already had a spinal fusion and a hip

scraping. She had Crohn's Disease (by the way, I have since learned that this is the only disease that the FDA says can be treated, legally, with DMSO/ MSM). She had an autoimmune disease called PSC (I have not fully researched it yet, but it sounds really bad), also, Fibromyalgia, Chronic Fatigue Syndrome, Lung Chlamydia , Q Fever and Lyme and Myopathy... but get this: She was due for a liver transplant at level 4 (I have to read up on this too as to levels). She was waiting for a liver transplant when she started this HL MSM Protocol (which I had not yet named then). The woman was dying... as she says and she was in terrible pain.
>
> Today, she still has symptoms of Lyme and Myopathy, but her liver and these other maladies are gone... zip! One year on this protocol (but not quite, she did not understand it yet) has done all of this. Her doctors claim that what has occurred is a virtual miracle... impossible!
>
> Now realize that one year ago, before I got really serious with my book, I had no idea about how the liver affects the joints or how MSM increases mitochondria/ telomere efficiency. This is still theoretical, of course, but the studies have been done and now we are seeing direct results that point the needle in the direction that my conclusions led me.
>
> Your constant, I don't want to call it wailing, but whatever, Duncan... regarding Glutathione Production and its affects is what led me to putting all of this together. I really believe that what we have here is a medical breakthrough in the making. Joint health was what got me started, but it's just a small piece of a much, much larger puzzle. I predict that what has occurred with Mona is just the beginning of a huge change in how medicine will be practiced worldwide once this gets out and it will all be centered on liver health. She has pointed the way to why, "It's the Liver Stupid."
>
> Kind Regards, HUUman

Hearing and Sight Degradation with Aging
3/8/13
Comments on article regarding Hearing:

The comment: "2. Human hearing starts to get worse at age 20. " is just plain wrong. If you include enough antioxidants in your diet, hearing stays the same. I proved that to my doctor 20 years ago. My hearing was fine from 20 Hz to 20Khz at age 69 and it is no worse now. My super high-end sound system has this range and I could hear all of those frequencies clearly before my son messed it up. It's now out of commission, but I am just too busy to fix it.

Vision is a different story: Presbyopia does seem to occur even with anti-oxidants. My vision went from 20/40 in high school to 20/15 at age 25 because I elected not to correct it with glasses. Today, I can read very small print in good light, but need 1.25 diopter readers in darker situations. Still, my distance vision is still 20/15 at age 71. However, what I am saying is that aging does affect eye strain or darkness vision just as we are told. But the real point is that it does not affect hearing to any degree if our diets are not SAD. I generally do fine reading my super high resolution (2560x1600) monitors set at max resolution without glasses most of the time.

Most of these homilies that are passed on today are based on what happens to a population of "idiots" who follow the government and mainstream recommendation of a SAD diet. It is hard to envision just how good things will get when we start getting back more results of the full Anti-Aging & Heath Program (AHP). My thought is that none of these old school standards will hold for even a minute. In fact, there is no doubt that my vision has improved, somewhat, in just the past year on the AHP. It has already dropped from about 2.5 diopters at its worst. Where it goes from here is anyone's guess. The fact is that we first age backwards on the AHP, so Anti-Aging is really a misnomer.

Regards, HUUman

Cooking Oils
Hi Paul:

So far, Macadamia Oil looks good as a cooking oil from all that I read. Red Palm Oil seems good and it adds alpha carotene to you diet which is nice (check it out) and it is cheaper to buy. Duncan likes MCP which is apparently benign in both directions as I read it, but very safe. Coconut oil is good, but as Dunc says, it imparts a taste... but that makes popcorn extra yummy. Not that I eat much of that either, but I still love it on a cold night watching March Madness which is coming soon.

Regards, HUUman

>However, when all is said and done, you just want to minimize
cooking with oils as much as possible for the most part.

Is Cancer Genetic or a Mitochondrial Problem
3/10/13

I have said herein that basically all disease is a Mitochondrial and a Glutamine metabolism problem and that the AHP can help stop them... with few exceptions. This is a ridiculous assumption in terms of how mainstream medicine looks at disease today, but I believe that it will prove out and the AHP is already giving us back strong evidence that this is true. My poster woman, Mona, is a living example. There are plenty of other examples, certainly, but not as eye opening.

.

Watch this video on Thomas Sefried, PhD, who agrees with what I determined in my research when writing this book.
http://articles.mercola.com/sites/articles/archive/2013/03/10/ketogenic-diet.aspx?e_cid=20130310_SNL_Art_1&utm_source=snl&utm_medium=email&utm_content=art1&utm_campaign=20130310

My premise: We can win this battle on cancer with HL MSM and the AHP and it is all about liver health... It simply does not take a team of twenty scientists and MD's to get this. Are these guys really talking? They are making this problem unnecessarily complex Like Dr. Trent in his video (first mention in ch 1, here), they clearly get half of the story and the rest is deemed a solid mystery. My theory is, if you make the problem complex, you end up missing the truth. The true is always really simple. Just open your eyes, listen, and pay attention and it will come. The source is not where you expect... the mind actually gets in the way.

Dr. Sefried gets into calorie restriction as a method of reducing the vacuolization of cancer. For healthy people, there are problems with stress in calorie restriction as Duncan has pointed out. However, in my opinion, the bottom line is that "some" calorie restriction is a good idea. Still, you just do not want to overdo it, either. It reduces glucose and increases ketones. Termed the Ketogenic Diet (R-KD) by Sefried, there has to be a healthy result, no matter whether we have contracted cancer or not.

Dr. Sefried sees calorie restriction, then, as a basis of treatment for brain cancer. IMHO, some calorie restriction should also be a method of keeping cancer from forming in the first place. Sefried, after the R-KD, treats the cancer with what he calls "natural drugs." He does not define the "drugs;" however, let's hope that they get the word on HL MSM and begin testing this Protocol. The same arguments should make some calorie restriction of value when using the AHP in treating and avoiding Type II Diabetes, most other disease, and health in general. For this reason, I no longer eat a true breakfast or lunch except for whey smoothies, etc.

Interestingly, Dr. Sefried is a proponent of the (LCD) Low Carb Diet, but also stresses reduced protein intake as does the Rosedale Diet. But here is where we see the rub: The AHP is about balancing a higher protein intake against the lower carb restriction and raising the fats in such a way as to not overload the liver. All the while, basically stopping toxin intakes as much as possible for the same reason. When this is all done, the liver is always the benefactor and the determining organ. The bottom line is, when the liver sings, the body is in consort. "It's the Liver Stupid."
Regards, HUUman

MSM for Liver Treatment
3/11/13
Hi Byron:

Thanks for your rely... so here we go, my friend:

You are undoubtedly correct if we were discussing mainstream doctors. Because it is certainly true that no mainstream doctors are treating Liver Disease with MSM. Furthermore, they would not even know how if they were. They'd be prescribing, maybe, 1gm caps of MSM, once or twice per day and getting "zero results." So far, it sounds like you'd be there too, but I am hoping that I can change that. So bear with me on this and I'll send you a copy of my book, so that you can follow along with the refs.

When I had essentially written my book, I was certain that I had put together the key to Liver Health... and, virtually, the true path to wellness, hence the title, "It's the Liver Stupid." However, only a week ago, I got solid proof that I was correct when Mona wrote me. Mona, my poster woman, was up for a liver transplant (level 4) just last year and as a result had multiple associated diseases. She got on the HL MSM Protocol and is now, well. It could have been a miracle, but I am inclined to believe that it was because of the protocol. I realize that I am biased, but you'd be hard pressed to come up with another objective conclusion. Her letter, on reading my book, started... "You will not believe this!" But I do and I actually anticipated it given my research. It had to happen and it will occur plenty more once this is out. No one yet gets it in or out of the mainstream.

Given her associated diseases, Mona was, as she says, "virtually dead," when I told her about it and she went on the HL MSM Protocol (then unnamed) last year. Both she and her doctors had given up. They had no answers. She was on 14 meds then.

It was a done deal: Today, along with a healthy, fully functioning liver, she no longer has Fibromyalgia, Chrones, PSC, a rare, but always fatal (till now), autoimmune disease affecting bile ducts, and a host of other problems. Mona still has Lymes Disease, but I actually believe that HL MSM will eventually get that too. When accompanied with some other protocols, I don't think that it has a chance. By the way, we know that no one treats Lyme with MSM either, right? There is no chance that It would work there either in small amounts, but the HL MSM should eventually get it.

No matter how many nice things you say about SOD, it is still second... and we agree. The "Glutathione Conjugation Pathway" (P35 in my book speaks this) is the key. So lets reorder your antioxidants properly: It should be: Glutathione (first & always), SOD, and Catalase, IMHO.

Now first off, what makes MSM/DMSO so effective for liver health? As the late, brilliant (if verbose), Dr. David Gregg (P46) points out in my book, and I believe that the man was ahead of his times on this: The "Oxygen Transport Pair" is not just an "antioxidant," but literally it shifts from a powerful oxidant to an antioxidant. This virtually supercharges the liver as MSM becomes DMSO in your body, thus retrieving and giving up the O back as needed (and intelligently by my premise). Combine this with the two M's with, "Methylation," and you have the key to inherent wellness at the mitochondrial level (P26... & well documented), then add Sulfur (P32, but you know this, mostly)... which turns out to, literally, be "liver food" and I am sure that Mona is now well for reasons that have been completely understood for quite some time.

When not used correctly, MSM is relatively ineffective or even unnoticed. It is just that no one ever researched or explained the HL MSM Protocol before at an objective level, even though many horse farmers seem to have been using it in their families for years. Keep this in mind, this is not my protocol, it was given me by a farmer. I just discovered from existing research how it works and why after using it empirically for many years. Mona got the word when I had no idea how it really works. Now I have a pretty good idea, but there is more work to do and people like you can help. I am just an architect, not a health professional, but I understand associations and these are pretty clear to me.

So I suggest that you reconsider what you have said, Byron, and get back with me. I would like to hear your opinion once you have.

Kind Regards, HUUman

>It is the liver stupid. The body has 3 main anti-oxidants SOD, Catalase and Glutathione not Glutathione, As far as I know no one is treating liver disease with MSM or CO Q. SOD, Catalase are part of a natural treatment of hepatitis and other inflammatory diseases of the liver and its used to bring high liver enzymes back into normal ranges by protecting the liver cells from oxidative damage. MSM is good for a lot of things but unless it has the right other co factors its not an antioxidant. Co Q is but it is not a master anti-oxidant of the body but it is important. SOD like glutathione creates trillions of free electrons for the body's use, unlike 100mg of COQ which is limited, but useful for certain things. Byron Jacobson<

MSM Questions
3/12/13
Hi Olushola:

You are not a new kid on the block here, Osh, so you really know much of this, but for the newer people I'll dig down some:

To answer your first question: it is both... people fail to "swish it around," thus negating some of the methylation and they do not get enough.

My book has been in the writing four years now and I have gotten nothing but good results, except as I mention. I actually have no idea how many have tried the protocol, but I have only heard two negatives in that time:

- One was Dee last year, who left the CCO Group in a huff because she and Dunc could not agree on anything. She claimed it upset her stomach. She as taking it, nevertheless, and had some positives to report that appear in the History section of my book. Honestly, MSM should never affect your digestion negatively. It makes no sense, but she said it did.

- The other was Mike... only a few weeks ago: He is apparently having a severe detox reaction that does not allow him to take enough to do any good. Those reports and exchanges appear in the .New Posts Section, called, " We Are Learning More- Recent Discussions" in my book.

As far as Mona goes, she is an exceptional discovery who just came out of the woodwork last week:

- Mona has had liver failure. How many people in our group in the last four years were there with liver failure (that is our only sample): Chances are very good that she was the only one.
- Her autoimmune disease is so rare that they only have diagnosed 30 others and they are all dead now according to Mona. So we can be certain that she is the only one who has ever tried MSM for that one.
- Chrone's disease: Actually the FDA has approved MSM/DMSO for treating it, So there is likely a huge sample for that, but who cares? That is no big discovery. It has been out there for years as Dr. Gregg says..
- The only things that Mona still has left are Lyme's Disease and a nasty bacteria that hangs out with it. That is another test that is ongoing. She and I are working on a transdermal Protocol to help her combat this. Plus, she was just not doing enough MSM, but she is all over that now. It may tale awhile, but I think that we will win this as I said in my previous post. Her oral supps were limited by her compromised digestion due to the host bacteria, so transdermal should get the job done.

Mona's opinion is that she improved "only" due to the HL MSM. She was nearly dead and no one had any answers. Her 14 meds were ineffective. Everyone, Mona included, had thrown up their hands.

We have loads of good reports beyond Mona, though, and many are documented in my book from P84 on. All of my local friends who have tried HL MSM have shown similar great responses, but none had anything like the problems of Mona.

My book is just not about supplements in general. There are tons of books on supplements. It is about learning how to find the ones that work best for you and an individual... To find your personal Anti-aging and Health Program (AHP). Its about learning how to discriminate between garbage and good... and how to raise your level of consciousness, so that you can become better at doing all of this. It is a "Do It Yourself Book on Health and Wellness." I just discovered during the deeper research that the Liver was the key to all of this. It set off a light that was previously dim. Duncan's work led me there though.

There are no drug reactions that I am aware of between MSM and drugs... probably because that would make no sense. MSM can be thought of as a part of your body... it occurs naturally, so why would it interact? If a drug were so bad as to intact with MSM, you definitely should not be taking it... Not that I could endorse very many drugs in the first place. Let's see, my list is short: Selgin and... what else? Hmm, that is it! Help me out, there must be at least one more.

By the way, most all diseases should respond extremely well to MSM, but there are exceptions. The disease, ALS (Lou Gehrig's) is one. But if you happen to have the onset of ALS, MSM will make you sicker... and fast.

I talk about the farmer in my book... really a marketer of horse MSM who also raised horses. I have sense heard of and talked to others who raise horses, since. We do not have many horse farms here, but a couple are actually clients of mine. They all know about MSM and what it does. One that I recently heard about uses it just like the guy who introduced me to MSM. So the idea is not new. What Farmers do is almost all totally empirical. They do not document much of anything. They just know what works, but they are a dying breed. Sadly, factory farms are running them out of business. I hope that we can turn this around with a higher demand for truly organic products which factory farms can't possibly meet. Interestingly, our local organic dairy farm sells more cheaply than the supermarket products.

Quantum... first, did you watch the videos? I do my best to describe quantum occurrences in the book. Certainly, an Empirical explanation fits your definition, Olush. We have two products that utilize quantum energy that we plan to market with this book. One is placed under the hood of your car and it increases mileage, torque and hp. Dyno tests prove a leap from 51 hp at the drive wheels to 101 hp and road tests has indicated over 124 mpg on a two mile level course at 50mph on several occasions using redundant gauges. We got 62 mpg at the Mercedes Benz test lab in Detroit for overall driving. However, under more normal driving conditions we only got 46mpg average over 40,000 of testing in a 2012 Jetta Sportswagon (36 mpg EPA highway rated). Still that is well over a 25% increase in mileage (overall mpg) and usable power is phenomenal. Do I completely understand what is happening? No. But I love it and it will carry over to our "HUUman" body cream that I am testing, I am sure. Once the website is up, I will sell the book, the car device and the cream on it. June is the target date, but this is a big task.

Thanks for asking... Love your questions.

Kind Regards, HUUman
>Are you referring to the dosage as an incorrect us of MSM?

>If you get heat because of your position on MSM, it might be due to your rather small sample size of 1, A sample size of 1 is statistically insignificant. Mona. Maybe Mona improved because of the interaction between the drugs she was taking and MSM. Your ebook does list some supplements that should be taken along with MSM. Also, it might be a good idea to document some of the work of your farmer friend. Finally you need to define how you are using the term "quantum" I know the term from my studies in quantum physics/mechanics, which seems to be different from your usage. It seems you might be using it to express an exponential jump in outcome, right? Olushola<

IBS Problem Solved
3/11/13
Hi Barb:

Good to hear your reply. You were lucky. At 67, you are in the crowd that needs plenty more, so do as much as possible... my friend Craig (your age), does 1 heaping tblspn every time he goes to the toilet. There just is not any amount that is too much and you virtually begin aging backwards from the inside out. Within about three months, it you will begin to show a much more youthful glow... hair shines, skin softens & wrinkles lessen, nails grow faster & stronger& you just de-age. There is NO downside. Your mitochondria are functioning like a 20 yr old. Its an almost a free deal too. One that you don't want to miss out on. $42 for 10# and it is without question the best nutrient in the world. Tell your friends... get this out there.

Regards... With Love, Jim

On 3/12/2013 6:49 PM, Barb wrote:
> kiss, I am actually 67.I think MSM is the major factor in curing my 20 year IBS problem.
> Thank you Jim
> From: Jim <Huuman60@comcast.net>
> To: box <bcox242@sbcglobal.net>
> Sent: Tue, March 12, 2013 1:55:56 PM
> Subject: Re: e-book
>
> Good. Thanks on the report.
>
> Now take about 20x that amount just to stay well Barb... following the HL MSM Protocol in my book. Its easy and safer than drinking four glasses of water. There is no sense taking chances.
>
> Please get back with me on your particulars. I am guessing that you were about 40 yrs old, since it gave you that benefit. If you had been 70, you would have probably gotten nothing unless you were some sort of super being.
>
> Any additional info on your healing is greatly appreciated.
>
> Please do not forward my book in whole or part w/o permission. If you see any typos or errors I'd appreciate your reporting them back to me. It still has one more edit.
>
> Kind Regards, Jim
>
>
>
> On 3/12/2013 2:36 PM, box wrote:
>>
>> Hi Jim,
>> Wow, just catching up on this forum. I would like to request your e-book.
>> I recently "cured" my IBS of 20 years, by taking 4-6g of MSM a day, plus olive leaf extract. I knew it was an anti-inflammatory, as I had used it for back problems in the past, but had forgotten all about MSM until 2 months ago.
>> Can't wait to read your book,
>> Thank you so much,
>> Barbara

Hi Barb:

You can rejuvenate at any age... but the glucosamine had nothing to do with it. The anti-fungal may have helped some, but just good to hear that you won.

On 3/12/2013 7:11 PM, barb wrote:
> I am pretty sure I am not a not a super being! I am pretty herbal literate, but that is sure a bunch of MSM. However, at a low amount, it sure slowed my IBS, combined with Glucosamine and antifungal. I am 67, so guess some sort of a super human being. Ha!` I have read you can rejuvenate until age 70. I have had IBS for 20 years, but have kept it under control, until the last 4 years, when all H broke loose. OK, where's the book? Want to read. If you have on a site, I am very willing to pay a fair amount to read.
> Thanks so much

Autism in Children
3/16/13
Dear new member to Sulfurstories,

I am so glad you joined. The way this list works is that when you join, if you will share your "sulfurstory" then listmates will know what to talk about to help you in your exploration of the good and bad effects you've seen with sulfur supplements or foods.

I set up this list many years ago when I was lecturing all over on the biological effects of sulfur, especially in neurodevelopment. I still lecture on sulfur, but mainly outside the US. My path has gotten a bit redirected in the last few years as I found a dietary treatment that does seem to help the sulfur pathways in a lot of people, and seemed to help more than what people saw through using sulfur supplements themselves. That research continues, but it has absorbed so much of my time that I slowed down the rate of bringing new sulfur literature to the list here, but there are also other members on the list who are into sulfur research and post it here.

The conversations on the list have slowed down considerably from years ago, but even so, there are many YEARS worth of great discussions in our archives, and the odds are we talked about what you are interested years ago, so please read the archives using the advanced search techniques you will find on the messages page. If you want to find posts referring to the medical literature, put PMID into your search criteria.

Also, please take a while to look through the links section and the files section for goodies that will give you a lot of needed background.

If you will post your questions/comments as soon as you can, that will spur people to post in answer to you, but most of the oldies on sulfurstories don't post anymore unless they see an opportunity to help YOU.

We look forward to getting to know you as your share your "sulfurstory".

With best wishes,

Susan Owens. list owner
Member of the Defeat Autism Now! thinktank
Head of the Autism Oxalate Project
Researcher in sulfur's role in biology with an emphasis on glyosaminoglycans

Comment on the above: Glyosaminoglycans are sulfur containing foods and supplements. Dr. Owens is an unqualified expert on the various chemical pathways (Transport Systems) in the body that allow us to take up and use sulfur beneficially. Her interest, as she notes, is in using it to defeat autism. This is just one of the many uses for HL MSM that must be explored more.

Jim Gates, the Owner of **Immusist** will tell you that autism is being defeated as we speak. Kids that could not even speak, take Immusist and begin speaking their first words and his formula can be used transdermally.

Now add the AHP, a powerful program that we are finding can heal virtually every disease except ALS. By the way, ALS treatment, so far, is thought to require lowered sulfur, but they could be wrong there too.

Getting Adequate Organic Sulfur/ GSH
3/19/13
Hi Group (addressing Teresa's letter below):

Dr. Berkson has about maybe 10% cure with ALA and Silmarin milk thistle, if he is lucky. Certainly, glutathione (GSH) is the avenue to full success, but he is lost in an alley somewhere. Selenium is getting closer, but he is still missing the master components.

Teresa is certainly correct that GSH is the key deficiency in Liver Disease. Furthermore, though not mentioned, a compromised Liver is the background problem for virtually every other disease on the planet. There are many things that can compromise the liver... Toxins, overeating generally, too much fat, just about anything ingested in excess can cause the liver to not function at its best and the result, if you are lucky, will just be a cold or flu, but if it is bad enough, cancer, MS, CFS, Fibromyalgia, heart disease, etc.

IV Glutathione is like using a sledge hammer to drive a 10 penny nail... there are just easier and more simple ways of accomplishing this. There is a better way, as I mention below.

NAC is a bad idea in any case as Duncan Crowe (for those unfamiliar, check out his website) will tell you. It will work temporarily, but it is just not the way home... and "never" take NAC, long term. Better yet, just avoid it, period, as a supplement. But, yes, it is used in emergencies by the mainstream.

Certainly, no Tylenol... and all of these, as well as just plain Aspirin... don't use them. They stress the liver badly. The Roman army was issued and used Aspirin in its natural form with its herbal complements... white willow bark (Salix Alba) for pain relief. It was also used by the Egyptians and Sumerians. The herbal form is safer, but still, don't overdo it. Bayer patented what they considered the active ingredient salicylic acid, and thus we are left with another unsafe, mainstream medicine that everyone believes is safe, just like Tylenol, etc. People actually do die from these "safe, over the counter, drugs", as you say, and the mainstream generally pins the cause on other things. There are fortunes tied to these drugs. And "too much" may actually be very little, when the liver is otherwise compromised. We play with our bodies, led by mainstream advice, as though they were chemistry labs... load up the test tube incorrectly and you can get results that no one could anticipate. You could end up in the ER, as Teresa suggests, or just come down with one of their induced diseases like a heart attack or cancer.

As Teresa suggests, we have the safe and healthy way out of this... and it's simple... undenatured Whey (Immunocal if you can afford it). Read my book: "It's the Liver Stupid." and learn how to actually get there. Also, read the book: "Glutathione GSH," by Dr. Jimmy Gutman... but Jimmy makes some mistakes... and he recommends NAC. Still, he points out the value of Immunocal in getting the GSH levels up to snuff and in getting the liver healthy. However, Jimmy also buys into the standard HDL/ LDL formula which needs improvement. I submit that the HDL/LDL ratio is still an important measurement, because when HDL levels are high, you have a healthy liver. Forget about the heart: When your liver is singing, your heart sings along.

Finally... and my real point here: Whey smoothies are great food, as Teresa suggests, but you can't depend on them for their organic sulfur. Organic sulfur is "virtually killed" when you process a smoothie. Without organic sulfur, you are missing a key factor in the undenatured whey. The HL MSM provides pure organic sulfur in spades, plus it adds a key ingredient to the punch... the attached methyl group that increases "Methylation." I cover these in my book with plenty of published references. Taken together with the Low Carb Diet, you have the Anti-Aging & Health Program.

Kind Regards, HUUman

>Great advice for those with liver disease or just wanting to improve your energy. Dr. Burt Berkson out of Las Cruces, NM has helped many with liver disease (especially Hep C) from around the world. He has a protocol to increase the glutathione using IV alpha lipoic acid for those patients with severe cirrhosis of the liver along with selenium (200 mcgs twice daily) and silymarin. When they leave his clinic they take 300 mgs of a high quality alpha lipoic acid twice a day plus the selenium and silymarin.

About 4 years ago I learned that low glutathione levels are common in liver disease. Shortly after I was able to use IV glutathione which revived me for one of my son's weddings. It was amazing the difference that it made in my energy. That's the beauty of glutathione. But IV's can get expensive and after doing more research I found that coffee enemas (thanks to Dr. Gerson) also increase glutathione in the liver by 600-700%. What you find is that there really is more than one way to get the job done. I also added the selenium and silymarin daily. Then I learned about the whey protein and that also helped. NAC is another supplement that works to increase the glutathione. And there are others too. The ER's around the country keep IV NAC in stock for the patients that come in with liver failure. Many end up with liver failure just from taking too much Tylenol.

Anyone dealing with liver disease would probably want to read up on Dr. Burt Berkson. These articles explain his protocol and how he left conventional medicine

For those with low energy you might want to try increasing your glutathione and see if that doesn't help. Like Duncan, I too find it to be the most important liver support and energy boost out there. Undenatured whey protein shakes with the selenium would be a great place to start.<

Thyroid Function
3/20/13

Brownstein's lecture on Hyperthyroidism Dr. Brownstein is a holistic doctor who specializes in this:
3.newsmax.com/newsletters/brownstein/thyroid_video/video3.cfm?s=al&promo_code=12D94-1

The question remains as to how the Liver contributes to the problems that Brownstein has outlined. The point I that he treats patients with thyroid hormones and this may not be necessary given the AHP. He finds that ADD/ ADDHD and brain dysfunction in children are often thyroid related. The T4 to T3 conversion may well be, basically. a liver problem. He also mentions that this is often an iodine deficiency which makes sense and he mentions also that soy should be avoided except in unfermented form as mentioned herein. Also, alpha carotene (as in red palm oil) is a great .supplement for avoiding this problem.

The Methyl Donors (a better understanding of MSM/DMSO and a comparison to TMG/DMG)

3/22/13

The question has been asked as to how MSM and TMG compare as methyl donors (CH3) for body processes. TMG has no sulfur component, so it cannot be compared as a sulfur contributor:

Impairment of methylation results in abnormal cell synthesis and elevated levels of homocysteine, a toxic amino acid and a serious health risk. Methylation can be inhibited by inadequately functioning key enzymes, excessive protein and fat intake, poor diet, inadequate intake of methyl groups, coffee, alcohol or by smoking. Therefore methyl donors are very important to our health.

MSM/DMSO (Dimethyl Sulfone & Dimethyl Sulfate) and TMG (betaine anhydrous), both act as a methyl donors which is important for a wide range of physiological reactions in the body. They work closely with other methyl donors. These include cofactors: choline, folic acid, vitamin B12 and SAMe, a precursor of carnitine synthesis. TMG is a natural organic that occurs in high concentrations in mollusks and crustaceans.

Both contribute to electron and methane transport, the biosynthesis of melatonin, and the synthesis of CoQ10.

TMG may help protect the liver against the effects of alcohol, possibly by stimulating the formation of SAMe. There are reports that point to this as a fairly certain possibility for high levels of MSM/DMSO, but there are other possibilities that could lead to these results, also. This factor would be very helpful for non-alcoholic forms of fatty liver or non-alcoholic steatosis.(write me for references on TMG... The MSM refs are in my book).

TMG: Can act as an adjuvant of the polymerase chain reaction process, and other DNA polymerase-based assays such as DNA sequencing. By an unknown mechanism, it aids in the prevention of secondary structures in the DNA molecules, and prevents problems associated with the amplification and sequencing of GC-rich regions. TMG makes guanosine and cytidine (strong binders) behave with thermodynamics similar to those of thymidine and adenosine (weak binders). (WIKI)

DMSO/MSM: When Thalidomide came into the news and because of its wide spread use in Europe and the connection to resulting deformities in babies, the FDA stopped everything being developed, including DMSO/MSM. It has been a struggle since, but it has been widely used for animals, especially horses. In 1978, the FDA approved DMSO as a prescriptive treatment for interstitial cystitis, an inflammatory condition of the bladder in women. However, 125 other countries have approved it in treatment of patients for such things as lupus, scleroderma, arthritis, and diabetic ulcerations. No one, so far, in the mainstream has ever used or tested it at high levels, so nothing is widely known about the HL MSM Protocol. It has been proven to cross the blood brain barrier in MRI studies.

In metabolism in the body, TMG converts to dimethylglycine (DMG) and contributes a single methyl group. Similarly, DMSO converts to MSM yielding Oxygen. DMG does not reconvert to TMG and here is where DMSO/MSM may be superior in this process.. MSM reconverts repeatedly according to the late Dr. David Gregg. The methylation process apparently occurs later but little is known about it..
TMG has also been suggested as a less expensive substitute for S-adenosyl-methionine (SAMe) for conditions for which SAMe is used, such as osteoarthritis and depression. However, there is no evidence to show that it is effective for these purposes.

TMG is extracted from sugar beets in the process of making sugar, but it occurs in other plants such as Spinach and also in meat. DMSO/MSM is made from wood (lignin) pulp in paper manufacturing and it occurs in primitive plants... read my book for more on it.

.

1. Barak AJ, Beckenhauer HC, Tuma DJ. Betaine, ethanol and the liver: a review. Alcohol. 1996;13:395-398.
2. Barak AJ, Beckenhauer HC, Junnila M, et al. Dietary betaine promotes generation of hepatic S-adenosylmethionine and protects the liver from ethanol-induced fatty infiltration. Alcohol Clinic Exp Res. 1993;17:552-555.
4. Mangoni AA, Jackson SH. Homocysteine and cardiovascular disease: current evidence and future prospects. Am J Med. 2002;112:556-565.
5. Kanbak G, Inal M, Baycu C. Ethanol-induced hepatotoxicity and protective effect of betaine. Cell Biochem Funct. 2001;19:281-285.
6. Abdelmalek MF, Angulo P, Jorgensen RA, et al. Betaine, a promising new agent for patients with nonalcoholic steatohepatitis: results of a pilot study. Am J Gastroenterol. 2001;96:2711-2717.

MSM (from a long-term user)

Bernard M. Cain who stopped by one day to the Coconut Oil Group:

(Please note that Bernard probably does not understand the HL MSM protocol, but he certainly knows about DMSO and MSM... I love this guy!)

I am 80 years old and I can read this without glasses. In 1950, I used DMSO, three level teaspoons, in 8 ounces of distilled water, drank it, snuffed it and rinsed my eyes, with an eye cup. The only problem was the smell. MSM does the same job with no smell or odor.

In the chemical trade they refer to it as 80 mesh. The powder form of MSM is usually because they have added a filler, to create more product, or in some cases to prevent clumping. Fillers are added to make pills and to put in capsule filling machines so as to prevent jamming or plugging. Depending on the use you can order it in 60-70-80-90-100 mesh. 80 mesh dissolves in warm water.

Dosage. Two level teaspoons in 6 ounces of water ***(Note that this not the HL MSM Protocol).*** For all external applications , rub on, or apply with a dropper.

The same amount is used for internal use, if you are over 35 add 30 drops Aerobic Oxygen. Take by the mouthful, and hold in the mouth as long as possible, then swallow. MSM is quickly absorbed in the mouth, then in the digestive system ***(Good job here Bernard)***. It only has a 3 or 4 hour life span, so it is necessary to take a mouthful every 90 minutes ***(Could be, but I have no way to test his assertion & twice/day works for me).*** Really good for most types of pain.

This dose: 2 level teaspoons in good drinking water is great for any eye problem. Use an eye cup do the right eye first, and snuff what is left in the eye cup up the right nostril. Then do the left eye, open your eyes so you can see the inside of the cup, then use what is in the eye cup to snuff up your nose. ***(If you try this & it works, let me know. MSM worked on my dog for osteoarthritis)***

Keep a bottle with a dropper handy to your night table, if your nose is blocked put a few drops up each nostril, within 5 minutes you are all clear.***(If you try this & it works, let me know)***

If you have an old dog with cataract blindness, apply a dropper full in each eye, two times a day, with in three weeks he will have his sight back. (Same dose 2 teaspoons (level) in 6 ounces of good drinking water) If you have questionable water get a gallon of distilled water.. If you have a dog that is lame behind (Hip dysplasia) mix same dose and put it out for drinking water.
(If you try this & it works, let me know. I wrapped MSM powder in hamburger for my dog & that worked well for osteoarthritis (OA) and hip dysplasia)

Anyone who suffers from Allergies-hay fever- ragweed do your eyes, and nose and you are good until noon. I have been using this in my eyes and nose for forty-five years... complete relief with in fifteen minutes.***(If you try this & it works, let me know. There are better ways of dealing with allergies as discussed earlier... And they are a liver problem so this will help)***

MSM and DMSO, *as mentioned above*, are by products of the pulp and paper industry, it is extracted during the paper making process, the raw materials are trees, rice stalks, grain stalks, squash and pumpkin stalks. DMSO is the cement that holds the plants together ***(interesting... Bernard knows some things as you can see, but does not go as far with it as Dr. Gregg did)***

Hi Osh & Group:

Synergistic Effects
3/23/13

It is just too soon to tell. We are still working this stuff out, really. But like you say: Look at your condition and treat it with all known protocols that you can afford. They all are synergistic in some way. If it is joints that you are treating: Hyaluronic acid, Egg shell membrane, Megahydrate, SAMe, Collagens & the various organic sulfur carriers, etc... They all make sense and I have always recommended this.

No sense in being a snob here. We want to get this healing job done and over, ASAP, then just get you to maintenance levels. I would far rather have a Group of totally well people than solid proof that MSM has done the trick.

For some of these conditions that no one had a clue HL MSM would effect just a year ago, like liver disease or autoimmune: Do your research and look at what others have used previously. Some have been very successful. However, I am seeing that w/o HL MSM these other supplements like ALA and TMG that others claim are so powerful will not put you over the top. With HL MSM, you just get well. That is our goal... wellness, not trying to make HL MSM a shining star.

By the way, thanks for the question, Osh... and that was a good one.

New Cream:
3/27/13

Actually, my friend Rich and I are developing some new protocols that will make MSM look like child's play (but MSM is still really good, mind you). I started this maybe two months ago and talked briefly here about it here and got some useful feedback, but now it is way over the top from where I started. My quantum body cream and spray now have my full attention.

I can actually program these transdermal delivery systems to effect a particular problem as they make you well overall and anti-age you. This works without changing the all-natural supplements that make them up. As you might guess, nothing like this has ever been done before. I got this idea from our work on quantized car engines, Dr. Mark Sircus' Trandermal Magnesium, and MSM/ DMSO. The "Quantum Orange Creame" (QOC) has various supplements that I have selected that are delivered through your skin in less than a minute with no residue. We will also make it up as "Quantum effect Transdermal Ozonated seawater" (Q-TO-U) body spray delivery system and we hope to have them marketable by the end of June. I am working on a website for all three of these right now. These Quantized Systems can quickly make a huge difference in our lives.

As you would guess, there is always a Herx reaction when you speed things up too fast. The idea is to program them to the individual and their condition. I may sell it in levels of power. So you buy a #1 and a #10 and you mix them yourself (when you are not treating a specific condition) to match the degree of Herx that you are willing to endure. The #1 really has very little if any Herx. Just today, we discovered that the highest quantum level (#10) rubbed onto your feet will cause a strong flu reaction and make you itch all over as you release toxins (no, it will not actually hurt you). The Herx from the #10 cream can last 8-12 hours before you get past it. Then you just feel great. Now that is power! Herx=Power and Power=Speed of Healing. MSM never did anything like that and its Herx is much lower.

The best part is that I easily control the power. This part is huge in itself. I can ask you, "How much herx are you willing to put up with till you get well?" And you say, well maybe a runny nose and a little itching... So I give you a #6 and mail it with some #1. When you try it, you drop back and flinch a bit and say, "Hmm... that's worse than I really want to handle." So you mix is some #1 and move on. Now when could you ever do anything like this with drugs? All the while you are anti-aging and getting well at "your" decided rate. Again, we are not discussing side effects (bad stuff), these are healing effects (good stuff). With Herx, you are paying the piper. There is no free lunch.

By next year, we may be making sick old 70 yr old codgers into 40 year olds. Four years ago, when I started working on our Car Device (Called a, "6 Pack"), I would simply have laughed in your face if you suggested that this could happen, but it is absolutely working.
Regards, HUUman

>Jim or anyone, in your ebook you do mention a number of companion nutrients. Please elaborate a bit on what nutrients are synergistic with MSM. I would expect that the choice of nutrients would depend on what one is trying to achieve.
Olushola<
Hi Duncan:

I like what Dr. David Gregg says once you wade through it. Even though he is not very clear on where he got his information or in expressing it, I see him as a brilliant and creative theorist who may be the Einstein of the "Oxygen Transport Pair." You say that you have read his report and dispute it, so be it. We disagree, but below is why he gets two Chapters in my book:
Gregg used DMSO empirically and marveled at how well it worked, then he went on and recommended it to his friends. Not surprisingly, it worked well for them. Up till then, he had done virtually the same thing that I have done with MSM. From here, he did some experiments with the "Pair" and discovered some of their basic qualities, such as" they both dissolve in oil and water (recently disputed by one of our Group here). Finally, his research led him to his theory. How he got to his Oxygen Transport Pair Theory is vague. The bottom line is, In this process, he created some theories that I really like and that make great sense to me. While he did not lay it out plainly, I break it down as follows and relate it, as best I can, to what you have posted previously:
The first part of Gregg's Theory is that DMSO is a powerful oxidant that when it gives up its oxygen molecule becomes MSM. You appear to agree with that and have stated as much.

The second is that once it becomes MSM, it is an antioxidant and behaves as one. This you seem to disagree with. You seem to agree that oxidized DMSO is MSM, but you believe that then it is virtually is useless. Gregg's Theory disputes what you say and I could never see the body rejecting MSM just because it has been "Transported." Why would it when it is so otherwise useful?

The third is that MSM is an anti-oxidant. We all agree as to how anti-oxidants work even if the whole theory is suspect to some mainstream scientists. An anti-oxidant picks up free oxygen radicals (O's if you will), thus keeping our bodies from rusting away in the act of aging. Gregg's Theory says the MSM, in picking up the O, becomes DMSO within the body and is thus a powerful oxidant again. You dispute that. I say it makes great sense and I accept it just as I do the above. What else could it become? And why would an intelligent body reject a great oxidant if it needed it? Do you further dispute that the body is a Quantum Machine that inherently "knows" things... and wants to be well... which is the basic premise of my book? Do you believe that in the absence of scientific proof, nothing can occur? I think not.

Gregg's Theory says that this "Oxygen Transport Pair" as he calls it, continues to reprocess itself. Thus, his conclusion is that DMSO=MSM, once ingested and the MSM>DMSO>MSM>etc. That makes sense to me. Why should one or the other just go away if they are there and an "intelligent body" "knows" that it can use them? You agree to part this, but dispute the obvious IMHO.

Gregg further explains (and I love this too), that all plants basically thrive on this pair in their internal natural processes much like the familiar photosynthesis. Also, that the smell of garlic associated with the "burning" of DMSO is why garlic is such a powerful healing herb. What he can't explain that concerns me, somewhat, is that the opposite smell from MSM>DMSO>MSM does not occur... of course, this is a good thing, but the smell does not occur and that is counter intuitive.

The bottom line is that Gregg and I have observed empirically that this "Pair" is probably the most powerful natural healing protocol on the planet. Thus, from my attitude.. it's THE most powerful: Because virtually every manmade drug creates added problems as it fixes or attempts to fix another. While Gregg may not have absolute proof, the fact is that all "Proofs" are still just "Theories." Most "Documented Proofs" associated with mainstream medicine are either in vitro or as tested on lab animals. Furthermore, we all know that both of these can be entirely bogus when they are related to people... And from there, they become test studies... which we know can be guided by money. So what you hang your hat on so often is really all "Just Theory." Therefore, Gregg's creativity, unless something else comes along that proves otherwise, in my mind, stands as a wonderful, creative explanation that absolutely holds water.

The two aspects of MSM that I was told about in the beginning, organic sulfur and methylation... are contained in the name, Methyl/ Sulfur/ Methane. You find no real documentation on them proving that they work, but you also dispute most of what Gregg says and all that I was told initially... but I did find and submit some documentation on from others in my book. This leaves MSM being simply an anti-oxidant (the one thing that you have given it, so far that I have seen) and that leaves it as being no better than any antioxidant out there. Thus, in your opinion, MSM is equal to, maybe, grape seed extract, for instance. Duncan, the results overwhelmingly prove that it is far better than these. Being even the world's best anti-oxidant supplement does not explain the reported first-hand results, so you have to find a higher "proof" my friend. Unless or until you do, Gregg's Theory stands as the one that reflects the results best. However, I also believe that MSM does what I was initially told it does by the salesman years ago... "methylate" while it provides useful organic sulfur (and to the liver). Only then, would the reported results make real sense, because, per the title of my book, and as I have discovered, "It's the Liver Stupid."

If MSM is a "stable construction molecule," how can it do anything but provide organic sulfur? That would make if a food like an apple, but with none of an apple's added benefits. How could it make people very ill from a Herx? How could it be an anti-oxidant as you earlier said it was? I simply agree with Gregg that what is reported is incorrect, so far. After all, as Byron Jefferson says: "MSM is known to do nothing for liver health," (So it follows: "Dismiss anything other than what I know and say as crap, outright"). Even this Alternative Medicine Community gets stuck and that is how human consciousness works. We are still people with very prominent egos and we love our ideas. We hate change... If our book says that green is orange, live with it, dude! You are color blind. There is no other explanation: green IS orange!

Folks, the so called "facts" simply do not jibe with the results... can we agree on that? MSM, till I got onto this HL MSM, was a very secondary treatment for joint ailments (only) and not even very good at it. Now we know that HL MSM does all kinds of crazy good things for our bodies. Duncan, please, can you buy the fact that "established proof" might be inaccurate? This is not a huge leap. Face it, HL MSM is not even on the same planet with established MSM uses, period. I figured that out years ago, but you are obviously not up to speed yet. I hope that you get it soon though.

Gregg says (restating the above) that: virtually no good laboratory studies or work have been done on this "Pair" and admits that plenty more needs to be done (my point). I have heard no comments on this from you except what you say below. You appear to think that what you have found is correct information, no matter what the first-hand reported results are. Also, you offer no references, not that I suspect that you are making this up, but you always give us copious references. Why is that?

Finally, you, Duncan, take all of the above a step further. You virtually call Gregg a lightweight and say that he is unscientific for theorizing what he can't outright prove. Such an argument would place most all of our revered scientists in the same boat. Einstein's E=MCC was not supposedly proven until the eclipse, which the scientists that I follow dispute as incorrect and proved nothing. Most of Nicola Tesla's work was never published even though he had most of it patented, he openly shunned scientific journals and groups "to their dismay." In my mind Tesla was absolutely THE preeminent scientist of the late 19th and early 20th C, bar none, even if the scientific community hated him. Tesla's work beat that of Edison by light years, IMHO... but what was Tesla really? A creative dreamer... and so was Gregg.

Kind Regards, HUUman

>If Dr. Gregg was correct he should be building proofs, not ignoring the proof that already contradicts his theory. We can only verify that Dr. Gregg's assumption quoted here had indeed been found to be incorrect in medical research, "When it is transported to the lungs where you have the highest oxidation potential, it could be oxidized back to DMSO, go back into solution, and not be released as dimethyl sulfide to the breath. Does this also happen? I believe it does.", before he proposed it. The unsupported assumption, methylation in the lungs, had already been shown to be a phase II liver function, not a lung function. The Ph.D. Dr. Gregg hasn't based his theory on any research but is just a feeling he gets. Shoddy work for a Ph.D.. the passage

also mentions oxidizing the stable construction molecule, MSM, also in the lungs, yet MSM is known to be a stable construction molecule.

all good, Duncan<

MSM... Why it Works So Well
3/30/13
Hi Jason:

Before I respond: I discovered last night that eating an organic lime slice, followed by swooshing MSM actually makes the MSM taste really good... forget neutral. I am now wondering why. It should also work with organic lemon. Lemon is next on my list to try. There may be a hidden reason as to why.

Duncan disagrees that the sulfur is bio-available for glutathione production, since it is not bound in a protein. This was just something that I assumed and made sense to me. If he is correct and you
are also on all counts, that would leave MSM as having no health benefits at all, other than aiding in collagen production, which everyone absolutely knows occurs... comments?

I agree that it is the results that count. I just happened to come across Gregg's explanation in my research and thought... well yes, this could be why this works so well.

I am wedded to nothing, not even the beloved "Scientific Method.'" So if what you suggest were Gregg's approach and that proves out, fine. Finding an explanation that matches observations is just fine, IMHO, if what you conclude is actually true. Who cares, really? Isn't that what drug companies do every day with the natural herbs that they reduce? Problem is that their products (drugs) generally turn out worse than the herbs and occasionally they go from being good for you to poisonous.

If it does "flip" and your body's innate defense intelligence "knows" how to take best advantage of that ability... there are two serendipitous events occurring, that would put MSM is a wheelhouse all
of its own, explain why, and bring up its value 5x as a defense mechanism. Also, the search would be virtually over. I would love to see it tested. Maybe if it were, we'd disprove it but accidentally
discover that real mechanism(s).

When you think about it, we have literally a trillion dollar drug class, Statins, out there today that, overall, probably do much more harm than good (and that appears to be the case). Sure they lower
overall serum cholesterol very effectively, but then when you look at what they do to the body, that is a bad thing rather than a good one in almost all cases. This as opposed to a naturally occurring organic chemical (MSM) that does a lot of good... with no good understanding of what it does. Now
this is a dilemma. We really must resolve this as quickly as possible so that we can maximize the benefits.

With the Low Carb Diet, when it first came out was not at all understood. It worked, so it was suggested. many took it to mean that they could just eat red meat and be healthy. Now we know a lot more. We know that it really is about limiting sugar and that carbs are sugar... big step forward. Now we know that green leafy veggies are really important and a whole lot more. We also know about the any health problems caused by sugar consumption and how hard it is on your system. Obviously, the drug industry has nothing to gain from this understanding, but we are forcing them to take notice as we reap the benefits of these new understandings.

Maybe when we isolate all of the mechanisms involved with MSM we can make a similar leap. My concern in doing this is that we have little to proceed with since it was never that important in the past and that, compared to drugs, no one really cares that we don't. They just keep treading on the same deficient past information. MSM is certainly not as important, overall, as Low Carb Diet, but it can be a milestone in health, nevertheless.

Thanks for jumping in Jason. By the way, have you read Dr. Catherine Shanahan's book, "Deep Nutrition?" It really is quite outstanding. Got my copy used on Amazon for one cent plus $3.95 S&H... what deal!

Kind Regards, HUUman

>I really don't see how MSM itself is antioxidant... not that it really matters. The sulfur MSM "donates" undoubtedly helps in glutathione production and utilization.

Once DMSO is converted into MSM, there would have to be some sort of energy process involved that changes MSM, allowing it once again to become oxygen reactive, one way or the other:<

MSM as an Antioxidant
3/31/13

Hi Jim:

Great (*on the lemon*), I'll have to try that. I'm back to taking about ½ cup of MSM daily, and I'm pretty used to the taste; the body adjusts to it when given time.

First, as a sulphur donor, MSM is an amazing detoxifier. It's well known as a liver detoxifier, but it's clear to me that it also detoxes at a cellular level. This effect can be charted and measured, and you
can predict how severe a detox reaction will be simply by charting pH.

I pulled out my antioxidant meter last night, and added MSM to antioxidant water with ORP of about -300. Adding MSM had no effect.

I added MSM to standard dead tap water, and was able to measure no antioxidant ability whatsoever. I'm not saying that something can't happen in-body to convert MSM into an antioxidant substance, I just haven't seen any indication of that whatsoever. In as of itself, MSM is clearly not directly antioxidant.

Sulfur, as I understand it, assists in the utilization of glutathione, not necessarily the production of glutathione. It makes sense to me to use the Whey for production, supplementation of even more glutathione if needed (ill people), along with PHOSPHATIDYLCHOLINE (and other high quality lipids for lipid replacement therapy), and use MSM to bring it
all together.

Here's a quote from Dr. Benjamin on MSM:

Glutathione has two different states within your body. There's reduced glutathione and oxidized glutathione. The ratio of those two signifies the overall oxidative status or the ability of your blood plasma to address oxidative stress. MSM improves that overall ratio. In other words, you have much more reduced glutathione that's able to deal with these free radicals. That's, I think, kind of the key of how MSM really – and DMSO also does the same thing – by controlling that oxidative stress or protecting from the oxidative damage can have these therapeutic [benefits]."

According to Dr. Mercola, MSM also changes cell permeability. This improves cell detoxification, and improves nutrient uptake (at a cellular level).

It doesn't make sense to me that MSM is antioxidant; I use the most expensive and most potent antioxidants in existence, and they do nothing similar to what MSM does.

MSM/sulphur's ability to strip acids from the body is astronomical. I used to give private tours to interested people studying balneology and balneotherapy; if I had a group of four or five people, I could do a stark experiment. There's a natural spring that I frequented that has naturally occurring, sulphur-rich antioxidant clay/mud water. I would use a clean sterling silver chain, and have each individual wrap it around a finger, and place the hand into this spring for about 30 seconds to a minute. Then, I'd have them remove it. I'd try to pick the "healthy athlete" of the group first. The silver would come back clean. Invariably, the ill individual would dip the hand into the water, and the silver would come out coal black.

This is the sulphur ripping the acidic waste out of the body. After three or four days of therapy in high quality water, along with some healthy eating modifications and plenty of drinking water, the experiment is repeated, and the silver would come back untarnished. Therefore, after three or four days of water therapies, the body was now able to keep up with the elimination of acidic waste.

I see MSM referred to as an antioxidant everywhere, but I think that it enables the body to protect itself against free radicals, rather than acting directly as an antioxidant. An article written in FASEB by Ohio State University states that there is really no evidence that MSM is an antioxidant; that this is an assumption based on observations of biological effect.

http://www.fasebj.org/cgi/content/meeting_abstract/22/1_MeetingAbstracts/445.8

I think the issue is important to discuss, because if I thought that MSM was an amazing antioxidant, then I might ignore other important issues concerning antioxidant therapy, and being in error for some people can cause problems...

I use ozone industrially (at 10% concentration) at least 3 times a week, and I have to have my antioxidant capabilities and my antioxidant nutritional levels optimal!

~Jason

To All:

Watch the Seneff Video>>>Problems with Statins

Some key points that Dr. Seneff makes: Low cholesterol sulfate accelerates aging.

- A woman normally has a level of 1.5 units of cholesterol sulfate in her blood. At term, the level is 24.

- Statins are a Class X Drug, meaning that they must not be taken by pregnant women (like thalidomide)(teratogenic risks) since cholesterol plays a key role in fetal development.
- Cholesterol is a major part of colostrum and breast milk.
- High LDL in mothers produces low LDL in babies.
- Raw cow's milk is high in cholesterol sulfate.
- Cholesterol Sulfate is both fat and water soluble.
- High sunlight decreases increases HDL and decreases heart attack rates as do D3 supplements.
- No one can get anything published that makes statins sound badly which tells us that ethics no longer have a place in the mainstream.
- The key to health is really Cholesterol Sulfate.
- Astaxanthin and Sulfur, both protect us from the harmful effects of UV and radiation damage in general. The combination is greater.
- Wearing sunscreen, then forgetting to, damages you skin more than never using it.
- ENOS (Endothelial Nitric Oxide Synthase), (common in red blood cells, platelet, artery walls and mast cells) is sulfate synthesizer under pathological conditions.
- We actually extract sulfur from sunlight and air using the ENOS by Seneff's Theory.
- Electron (earlier interview) and sulfur storages are our primary storages.
- When the above are depleted, you take the sulfur from muscle and cause Muscle Wasting, MS, Arthritis, Diabetes, and Alzheimer's Disease.
- Low Cholesterol, then, causes the above diseases, because it is Cholesterol Sulfate that serves as the food for this.
- 25% of Americans over the age of 45 take Statin Drugs and few (if any) are warned of its health risks.
- An analysis looked at 72 trials which together involved close to 160,000 patients. It found that statin treatments significantly increased the rate of diabetes and liver damage.
- Statin drugs also interfere with other biological functions, including an early step in the Mevalonate Pathway, the central pathway for the steroid management in your body. So statins bring down the following levels:

- All sterols, including cholesterol and vitamin D (which is produced from cholesterol in your skin).
- All sex hormones.
- Cortisone.
- The dolichols, which are involved in keeping the membranes inside your cells healthy.
- Coenzyme Q10 (CoQ10), which is critical to the energy generation in the Krebs cycle in the cell.

Your body easily converts Organic Sulfur to Cholesterol Sulfate as long as you have adequate Sulfur and Saturated Fat in your system. So the bottom line is: Take your HL MSM and Undenatured Whey, do the Low Carb Diet, and avoid Statin Drugs like the plague.

Kind Regards, HUUman

Acid & Oral Health
4/2/13
Hi Tre & Craig:

First off... Sorry Tre. That one whizzed right by me. I should have known better. Your point is well taken. However, I do get worried when someone could get the wrong "part of the story." No doubt, If you sucked on limes or lemons very long without something to neutralize them, you would get cavities. Also, If you went around eating lemons and limes without being mindful of their acidity, it would probably not turn out well.

My first-hand story was that those oral C's dissolve in your mouth tabs caused a huge cavity years ago and one bad experience is plenty. It was actually a quick cause & effect. I had not had a cavity in twenty years prior to that... & there it was. That cavity showed up in an x-ray and I had to read it to believe it. Today, I never go to dentists anymore... not in ten years since I argued with my dentist over mercury fillings. He has changed to plastic now, but I discovered that I don't need cleanings if my hygiene levels are high enough. Still, If a filling breaks, he sees me.

My oral hygiene today is pretty extreme today, though, compared to when I relied on dentists. I use those lovely plastic flossing toothpicks after eating anything and I waterpic fairly often to boot, plus, I use the Sonicare System. In those days, the oral plaque would buildup to the point that I had to have a cleaning every three months or I could not even floss without breaking the string. Now that was bad! It was probably also related to my high fruit and sugar diet... I ate a slice of pie every day then at lunch.

To answer the real question: I swish immediately after the lime. I would think that just swishing with water and MSM after the lime or lemon would be neutralizing enough. Water, alone, is quite neutralizing as you know... and swishing water and MSM should help a lot. It would be a sort of sand blasting effect, LOL.

To the above I add: Considering how fast the above cavity problem occurred, my teeth would be ruined fast enough if I were wrong. We must be mindful that this IS an experiment, as Tre is seeing here. We are really pioneers. Certainly, we do not want to end up with a mouthful of cavities just to alleviate a small issue, but the lime/ lemon really was a nice find, I think, for those who are not past the taste issue. And I actually do "Like" the taste on MSM following a lime slice. Also, there are likely subtle reasons for this. Our taste buds work for a reason, Other than sugar, if something appeals to your taste, listen! Our bodies are intelligent. They know what is good for them.

As some of you know, Jason and I have been eating organic lemons for quite sometime now (me in smoothies) with no apparent ill effects. As Jason reports, they introduce awesome health benefits once they are in your body. To that I add the their skin is much more nutritious (full of C) than the inner fruit. Never heard about the seeds, but they are probably chock full of nutrients also. The rest of the smoothie is probably enough to neutralize the acid in that case... especially the cream and whey. Also, we know that the net effect is basic once in your body. However, it still does not register as to why, with me. I believe it, but it does not make sense.

Furthermore, If you research it, both lemons and limes are low in sugar and limes are the lowest of all citrus. Grapefruits are also quite good in terms of nutrition as Jon Barron recently reported. The red is better than yellow, but they are both good... and better than oranges. We should still consider them as desserts rather than staples, nevertheless. Citrus fruits offer no good fats and all provide some sugar. There is no such thing as a beneficial dietary sugar (or Carb) as Dr. Catherine Shanahan says in her book, "Deep Nutrition"... the book that has become my bible for the Low Carb Diet. Cate gets it on that and tells it as well as anyone could. As Duncan says, sugar ages you and, as Cate says, we have to count sugar rather than calories if we are really serious about our AHP (Anti-aging and Health Program).

Kind Regards, HUUman

>Hi Jim,
I thought that maybe Tre was referring to the lime juice you were using to hide the wonderful taste of the MSM. That would've been my question too.
Regards,
Craig<

>Hi huuman. I was interested in the acid complexity given the lime that you ate. How long after eating the lime did you eat the msm? Limes and lemons are acid and with msm as a component that might hurt teeth. Lymeover. Ps thank for your explanation but now you know "the rest of the story" as Paul Harvey would have said.<

Go at Your Own Pace Please
4/5/13

Hi Peter & All:

>Hi HUUman: I started (MSM) slowly and built up over time so far have taken over 1kg and am only up to 6tsp a day. So yes, taking it slowly, and yes MSM gives me the runs and nausea occasionally, the skin rash started before msm. It started with the whey, thought the msm would clear it, but not yet. Good pain relief, but still need the morphine when it gets too bad, i.e., when I overdo it. Many Thanks All Pete<

Pete, you sound as though you are doing fine, but for others, think of it like this: You got this far without MSM and you've survived. You have no reason to rush. If these things become a problem, simply cut back. If you have health problems, there is never a reason to create more. In every case, time is illusion and you have an infinite amount of it. This is not a race to wellness. Let me explain this further in detail (and I do have a point here for you, Pete):

Your body "knows" how much of anything that it needs. Your mind is far less competent in that regard. As I say in my book, your body is a "Quantum Machine." That is something that we have learned at the expense of Newtonian Science that is quite literally, by his standards, impossible... But we know from many results. If we are paying attention, it is all of reality. That is, it is all around you and occurring within you. On one hand, your mind is a reasoning device. We need rationality to an extent. To be "whole" we must have both.

I had an extensive dream lesson on this just last night that I am passing on to you. Take it for what you will. Some will get it and others will dismiss it. These things work according to what we have earned spiritually and I respect both opinions.

Newtonian Science is rationality taken to one extreme. Magic is Quantum Science taken to the other extreme. Both are problems spiritually. Our goal should be to balance the two, really. Black Magic is an attempt to use Quantum Science to control other people or cause harm in some way. We must avoid all extremes as spiritual beings. Health and wellness occur when all of these are within the middle ground... the high road that we learn to allow to occur quite naturally over time.

If you have not watched the Dr. Trent Video please do so. This was not specifically in my dream, but one key point in it initiated some of this topic.

The point that is on the forefront is this: In Trent's video, he mentions that a patient had a blood transfusion and actually picked up memory from the person that they had received the transfusion from. We also hear similar stories from transplant patients. They suddenly can play piano or enjoy classical music when, in the past, they where tone deaf or hated it. Or, after the transplant, they are more forgiving or they are more inspired by art. That is, they have gained a degree of wisdom from the person from whom they received the organ (or blood in Trent's case). So what is going on? And why does the Newtonian World ignore this information when it is right there in their face?

Dropping back many thousand years, as hunter gatherers, this ability and knowledge was critical for survival. Now going from my dream: We always ate the organ meat first, because we had to know what these animals were doing, how they thought, and where they were going. In those days, we were not using high powered rifles and we needed their flesh to survive. Using a bow and arrow or spear, we required a great deal of additional information to defeat these intelligent and healthy animals that allowed us to survive. This degree of knowledge was critical. So we ate and learned. Often, we ate this organ meat uncooked, so that nothing was lost in the electron transfer. Quite possibly, the animal's heart was still beating when we ate it... sound gruesome to you? Well, it was not IMHO. We are allowed to survive and it is our right as spiritual beings. Only waste is wrong. We must be frugal in every way that we can, even today. We are here today because our ancestors understood that these characteristics were passed on through living tissue and that all living tissue carries this "quality of knowledge" to some degree.

Today, Newtonian Science has raised a crop of virtual idiots in the above regard. They are blind to this greatest of all natural resources... the Quantum Field. A few have gotten it recently, but we are really just beginning, compared to what our ancestors knew. They preceded Newton and they were better off for it, even though our ancestors were quite rational in most ways. Yes, today, we have cars, machines, modern medicine and drugs... and oil, but they had infinite knowledge at their finger tips. We consider ourselves as the superior winners. However, until we actually learn what they knew and unlearn what we grasp so tightly... They remain the all-knowing and we, indeed, remain the extremely rational, out of balance, "modern" losers.

On the Quantum Level, many rational impossibilities become possible. The Newtonians consider them as illusions of the mind or worse, magic. However, when you actually see it operating as Dr. Trent and I have, you simply know that these are factual events and are being ignored by overzealous rationality. Our bodies, as I say in my book, are powerful "Quantum Machines." They run on electricity and virtually every cell in your body is intelligent. Furthermore, your body knows itself completely and communicates between cells instantaneously... there is no linear time on the quantum planes. These, so called cells, just "Know." In our dreams, we "know" also. We actually have access to planes beyond the quantum level and I mention this briefly in my book, but few have learned how to access them. All are denied access to these higher worlds until they have learned to embrace love without question. Otherwise, these are an impossibilities. Then, they are not really places, either, in the sense of what we think of as places... but there are no thoughts there either, LOL.

Getting back to reality as we know it: When you question the Quantum Force, you relegate it to less than its potential. When you embrace it... you allow it... and bring it to the forefront... that has to be your goal if you are looking for its full effect. That is, as a quantum being, you have immense control over this overall force. When Newtonian Scientists look into it with their negativity and doubt, they see, literally, zilch, for this reason. Also, where you live on earth affects the Quantum Force. There are places where it simply works better. That is, there are what some call, "Power Points" where this force is more fluid. Some are quite intense and others less so. Therefore, people and life in these places Thrive at a higher levels However, you as a Quantum Being can create and develop these places. You can create them in your home, for instance and you can invigorate you body, of course, by making it flow through you more potently.

So in healing yourself and in anti-aging, if you are going to gain the maximum potential of this great force (that many attribute to God and worship in fear... which is silly), you have to embrace it. By the way, God is far, far greater and infinitely less accessible, but this IS a part of It just like everything else. So again, listen to your body. It simply "knows" how to stay well.

So the final point, Pete (and this really was where I was headed, LOL): If you get diarrhea, then, it is because your body saw that you were overdoing something and is expelling it before it can make you ill. Embrace that ability also and be thankful that you are what you are. It is a gift extolled upon you as Soul. If others, your doctor, your neighbor, or your wife, doubts this capability, disregard their ignorance and let them learn at their own pace. They are here to learn just as we are. We all have to get it eventually. Our job is not to change the world, but to embrace it with love. It is, in fact, "perfect"... We see it as imperfect because it does not work as our little minds expect. In Fact, we are all just where we need to be and learning at the level that we need to learn. There is never a need to press these issues, but we, absolutely, can learn how to enhance them.

Kind Regards, HUUman

Frequency and Vibration
4/7/13
Hi Jason (& All):

I have to say, I just love reading your posts, dude. You are so tuned in that it raises my awareness level just reading them... and I thank you for this one, especially. In reading them, I begin to wonder what the population as a whole would be like if they were all as tuned into their bodies and as aware as you are, but we have a long way to go and as I suggest below, there is always plenty more.

I hope that you watched Dan Nelson's video... the guy that Mary Jane referenced to us. I have watched it at least three times. He has the one that I post and another, more watered down version, that comes in four parts for beginners. But the first one is just amazing. I am in the process of getting in touch with him. He can help us explain some of the outstanding results (actually impossible in Newtonian terms) of our car experiments... which can't be that far from what we can get with people as we apply them. While it will take a few years to see how they translate to people, cars, being much more simple machines, translate results much more quickly,

Dan Nelson's results with water make all of the sense in the world to me, given what I already knew and have seen. Too bad that I can't make a Time Reverse Laser and make my own "Nelson" water, but Dan's water appears to be just what he says that it is and works as he describes.

No question here that "everything" is frequency and vibration and that he can actually replicate virtually all food using water. I have not written about it here previously, but yes, there are people on earth, referred to as "God Eaters" by those who are aware of them. They actually subsist on vibration alone. Their diets consist of changing frequencies, nothing else. We in the "modern world" (LOL) have a long way to go compared to where we are now to get there, but this is absolutely the ultimate form of subsistence. When you live on vibratory rates alone rather than unreliable food like we do, you get no impurities and, theoretically, you have virtually no lifespan and never age... and, no, they don't age.

Even with food getting in the way, our lifespans can increase dramatically and we can achieve the same lifespans as those reported in the Old Testament if we get our acts together... and as Dan suggests. This can occur even without the Xenon Gas that he refers to in his video simply by replicating those frequencies and there are several ways to get there. But again that is not the ultimate solution either. I love having his video known, because it allows me to tell you more than I otherwise could. I am only allowed to say so much. But of course, only a few can even hear these things anyway, so it does not matter. His video helps fill in the gaps between the physical and spiritual. it is very valuable to those who are ready to hear the subtle truths.

Dan tells us that food supplements get in the way of his water because they leave "ash" and this is true also. His water is a step away from the above ultimate delivery system and if he learns how to do it correctly he can learn how to avoid food entirely, but this is still not equal to what the God Eaters do. Once you achieve this level, you are no longer so confined to your physical body... but no one really is anyway. This is just an illusion and Dan kinda gets that and hunts around about it in his talk. Their demonstration of praying over food, by the way, is very rudimentary. There is no need to do this. once you have achieved a level of knowingness, these things occur automatically. That is, once you have achieved a high enough spiritual level, you can simply "know" that your food is good and it will be. Again, vibration is everything.

I am suggesting that this is where Dan's work is headed, but he does not yet know that, yet. He is closing in, though, with his nano particle water. Right now, I just am hoping that I can hook up with him at some point. One of my colleagues actually lives very near him. The three of us need to talk. I wish that you could be in on that too, Jason. I am sure that you would have some valuable input.
Kind Regards, HUUman

>LAST, I've had to stop taking MSM for the time being. The more I upped my dose, the worse my fatigue became (becoming worse over, what? a 90 day period?). I wondered why, so I began doing some research, and discovered that any sulfur based compound can be a problem for those suffering from hypothyroidism.

So I checked my basal temperature, and it was 95.5. I was shocked, because I thought I'd solved my thyroid problem. I stopped taking the MSM, and within 12 hours, my temperature returned to normal, and my energy levels dramatically improved.

I'm very grateful I figured this out, because I have a massive amount of manual labor to do next week!<

Boosting Cellular levels of Glutathione
4/8/13

Hi All:

Byron Richards offers this report on Fisatin, which I have supplemented for quite sometime for its reported ability to help sharpen mental processes. The interesting thing here to me is that his report claims it "boosts the function of cellular levels of glutathione" and I was unaware of that. Of equal interest though is it's ability to reduce "inflammation" which first got my attention. Given that I started it with several other "brain" supplements, I can't offer any first-hand reports on its effectiveness.

Snip from Richards>

"Scientists continue to document the broad neuro-protective and neuro-rejuvenating properties of Fisatin. One of the latest animal studies shows that Fisatin has the ability to boost the natural production of serotonin and norepinephrine, making it a mood boosting nutrient with anti-depressant properties.

Fisatin performs multiple roles in helping your brain. One is that it has antioxidant properties. Another is that it boosts the function of cellular levels of glutathione, the most important antioxidant system in any cell. It also reduces glial cell inflammation, helps stimulate the formation of new nerve cells, and has cognition-enhancing properties. A recent study shows that Fisatin reduces inflammation of the blood brain barrier, helping provide stability to this important line of brain defense.

With so many potential toxic insults to our brains lurking everywhere, from environmental pollution to germ toxins, digestive toxins and stress in general, it only makes sense to bolster your brain's vital protection systems."
Looking in Wiki, I see that it " has been shown to alleviate aging effects in certain organisms, such as the yeast S. cerevisiae, the nematode C. elegans and the fruit fly Drosophila melanogaster" However, these effects have not been replicated in humans. Also, it is flavinoids related to quercetin and a potent antioxidant that can alter signaling pathways leading to increased expression of several protective and antioxidative genes... And it lowers levels of oxidative stress markers.

"Fisatin, among other Flavinoids, was found to be a strong topoisomerase inhibitor. These are enzymes that control the changes in DNA structure by catalyzing the breaking and rejoining of the phosphodiester backbone of DNA strands during the normal cell cycle. This effect may be anticarcinogenic. But it also has one carcinogenic potential: As aDNA topoisomerase inhibitor, Fisatin and other Flavinoids may increase risk of acute myeloid leukemia, a rare disease and that risk is only for infants.

In recent years, topoisomerases have become popular targets for cancer chemotherapy treatments. It is thought that topoisomerase inhibitors block the ligation (the covalent linking of two ends of DNA molecules using DNA ligase or RNA molecules with RNA ligase (ATP)) step of the cell cycle, generating single and double stranded breaks that harm the integrity of the genome. Introduction of these breaks subsequently leads to apoptosis and cell death." Topoisomerase inhibitors can also function as antibacterial agents.

Fisatin can be found in a wide variety of trees and plants, naturally, including strawberries and, apples, and grapes even though WIKI calls it a "drug" and I believe that it should be of

interest to this group. It is available from most supplement stores online including Swanson and is not expensive.
Kind Regards, HUUman

MSM Does NOT Harm Teeth
4/12/13
Hi All:

I would say that DMSO is probably one of the most effective supplements in the world, from what I have read, when used either orally or transdermally. But there are obvious problems with DMSO that MSM avoids, even though the two turn out to be exactly equal, internally, as I explain below.

First, on the plus side, DMSO is a quick and efficient oxidizer... meaning that it delivers free oxygen molecules to any site on your body and that your body knows how to deal it with it in fighting a multitude of problems. So if your knee is hurting, rub some on and the body immediately will use it without any intermediate digestive paths. Orally, it is no different really. If you have a mouth infection, it goes straight to the source and it is absorbed, there it releases its oxygen bond and kills infections on the spot... but then, some of it continues on to the rest of your digestive system where it can do its magic there.

In my book, I repeat Dr. David Gregg, PhD's saying that DMSO and MSM are an "Oxygen Transport Pair" and one becomes the other when ingested. So they would be a long-term oxidant (powerful destroyer) and anti-oxidant (free radical bonder, if you will) that continues transferring back and forth for hours as needed. Duncan disputes this based on his readings, but this would begin to explain the reason that MSM is the powerful healer that we report. Since Dr. Gregg died recently, we can't be sure where he got this information. However, he obviously knew a great deal more than anyone else that I have encountered about DMSO/MSM, so I believe that he was onto something. I am inclined to accept his theory till a better one is proven, but I also know that this is not enough to explain all that is occurring. There is more...

Actually, given what the astrophysicist Dan Nelson writes, MSM/DMSO could be far more powerful than just an oxidizer/ antioxidant as I suggest above. It could actually be involved in the transmutation of elements in its pathway. That would account for the strong metallic taste that one gets when they first begin using the HL MSM protocol. I recall it occurring often and decided then that it was just leaching mercury from fillings. However, on thinking about it, that does not make sense, since that taste stopped occurring some time ago and my fillings are still intact. Dan reports it with his water and I know exactly what he is talking about. So the two may be one thing.

This is pretty far out, but, in fact, Pons & Fleischman did encounter "Cold Fusion" and their experiments have been replicated at least a thousand times in the past twenty-five years (despite what Dan knows in his "White Paper." Additionally, there have been many books written since, on the low temperature transmutation of elements by noted scientists and even our own Dept. of the Navy agrees that it occurs and has replicated it.

Dan Nelson, who I just got onto recently, discusses the above event occurring within the body, and logically, when he discusses the conversion of magnesium to potassium in cell transport. Also, outside the body, he discusses the transmutation of water to gold. This transmutation process could be just one of the elements that make us the quantum machine that I discuss in my book. Actually, I believe there are plenty more and this is the tip of the iceberg, but there is much to be done to prove any of this. However, if it is true, Sulfur may be transmuting into any number of helpful elements as required.
So the above conclusion, though currently just theory, could some day be proven to be fact. Furthermore, it would explain the workings and results of using the "MSM/DMSO Pair" better than any other theory. But we are just beginning and it will take years before we actually know what is going on. In the meantime, just use them and be aware that you are using a very powerful protocol that is showing dramatic results first-hand, where it counts.

Now, for the negatives:

If you are particularly ill, it is thought that DMSO causes your body to leach off contaminants in much the same way that garlic does, but in doing that, you smell exactly like garlic. In fact as Gregg points out, the greater the health problem, the worse the smell. If you have ever been around someone who takes raw garlic cloves orally, you know the problem. They just smell horribly, but they are getting well. I believe that Bob Beck was wrong in his general assessment of garlic, but he may have been dead accurate in advising not to use it with his device. The two may be too strong together for the body to deal with at once and that would explain his bias. By the way, if you are wondering I have used both the HL MSM and the Sota device concurrently with no ill effects, but never DMSO with it. So that could still be a concern.

Next, DMSO can be just as dangerous as it is effective if you happen to have contaminants on your skin, because eye shadow or other beauty "aids" can be deliver them straight into your system and long term and could possible be toxic. So it requires absolute care in using it. Therefore, if you are a mechanic and have dirty nails and rub in DMSO into a knee joint, you could be inadvertently delivering dirty auto grease into your system and that, long term, could be life threatening.

Finally. I have read literally reams of papers on both DMSO and MSM in my research and I have never seen anything credible that would lead me to the conclusion that either MSM or DMSO could harm tooth enamel. However, I did have a person inquire once where he had a new filling fall out after starting the HL MSM protocol. It did worry me, because we know that MSM picks up heavy metals and expels them as discussed above... which is a good thing long-term. Then I learned that he had lost a "plastic" filling. There is no evidence out there that I am aware of that would suggest that either DMSO or MSM affect teeth enamel or plastic fillings. My assessment is that this man just had a bad filling... bad dental work. Unless it happens more than once, there is no reason to look further.

The bottom line here, then, is: Please don't put out "off the wall comments" that you can't back up. Think about it. People forget or don't have the time to research things, but they recall bits and pieces of these comments and might miss out on a helpful cure as a result. In that case, you have done unjust harm even if you had no such intention.

Kind Regards, HUUman

The Body Electric & More
5/3/13
Hi Vicki & RJ:

Beginning in about 1976, when I first transcended my body consciously and got to the realm of pure light and sound (which, as most religions teach, is accompanied with an overwhelming feeling of greater love), I realized that the brain, itself, is just a receiver and transmitter... not a place where things really get done as the mainstream believes. Furthermore, you do not have to be in a near-death state to go there... or on drugs (and I was neither). The mainstream is just dead wrong on this stuff as are religions. Since that time, I have accomplished this a few times over the years, but I can't do it just because I want to. What I have come to discover is that there has to be a "need" for this to happen for the greater good. Maybe that first experience was just for me to "get" it. After that, though, they are not just gifts for me to learn from.

Recently, I began reading Astrophysicist, Dan Nelson's White Paper and watched most of his videos. Dan basically agrees with and repeats what I discovered independently, so it is nice to have some confirmation from someone who has a more scientific bent than I.

So what does this mean? Dan and I are in agreement that the brain is "way overrated," but when there is a fault with it, it is simply not able to pull in the stored information from our various auras that surround our bodies. Yes, the brain is a delicate transmitter and various things can affect it negatively, but at the same time, we are far more than just this transmitter/ receiver (or our physical body). In fact, as science is just beginning to understand, our whole body transmits and receives information from our outer auras. That is, our organs can learn. When someone gets a heart transplant, they inherit information from the person who donated that organ (and many have reported this). Even our individual cells do that... Further up the line, so does our Heart Center itself, a powerful chakra, or energy transmission point long recognized by Eastern Religions, but basically discredited by Christianity.

So as Dr. Robert O. Becker tells us in his book, "The Body Electric," our bodies are basically electrical transmission centers. When we die, we shut down all at once, internally. It is an overall agreement, not a slow turnoff. We just leave. The cells stop communicating instantaneously in this agreement. With this understanding, Bob Beck's electrical devices do work in cleaning up the flows of electricity in our brains (and throughout our body as Vicky reports). When the various fields are opened and transmitting better we feel a lightness because we are in better touch with our "real" selves.

This phenomena is not just a mechanical mechanism. As Dan Nelson says, This is absolutely a "Quantum Effect." This is why I say in my book, "It's the Liver Stupid," that we are Quantum Machines. We basically underrate ourselves and see ourselves as the lowest part... our physical body. It is the only part that most can see (some see Auras) and we love physical verification. The same effects that Vicky reports with her magnetic pulsar can be achieved in spades by simply doing contemplative exercises where we envision ourselves as who we really are... Soul is one term for that Outer Electrical Aura. Therefore, we are that "Aura" that Dan talks about. This is our most important part and, as that part, we never die. What we do, when our body quits, is lose some of the lower functions of that Aura (the astral, causal, and mental layers)... so we forget, to a degree, who we are. Just to really mess up the little box that we put ourselves in, some don't forget the whole story. For this reason, the idea of reincarnation persists, because, guess what? It is true and this is how it works. The mystery is that we are so resistant to these truths and are so quick to discredit them. The evidence is everywhere and it really is a much more comfortable story than just a physical box that dies and goes to sleep or just disappears into thin air.
Kind Regards, HUUman

Hi Vicki,

Can you explain how you positioned it on your head, "for example: moving around? holding in a position for How long? Looks like the Neuro Star has a specific area in the top front for depression?
Just curious, I don't want to screw my brain up more? By the way does Russ from SOTA say it is okay for head applications?

RJ

Vicky wrote:
> I contacted this company because I'm always looking at new things. Unfortunately not transparent. Asked for basic output specs - a reasonable request to know what is going into one's brain. They wouldn't answer me, said that I had to already own one, or have, in practice, actively seeked this technology. No specs on line. The only thing I was able to find out was that it output 135 v/m.
>
> The red flag for me was when they list the most common side effect with this Neuro Star is "pain or discomfort at the treatment site." They wouldn't explain that either. I asked if that was just because of the pressure against the skin, or what and they said I had to own one or have a practice actively looking for this technology. Yowzers. I have never heard of a therapeutic magnetic device of any kind having "pain or discomfort at the treatment site." All in all, I was fairly displeased with the lack of transparency for non-proprietary info.
>
> Okay, that rant aside, I can tell you that I've been looking into transcranial magnetic therapy for a few years because, one of the common "side effects" that I hear from the use of the Beck magnetic pulser, is a feeling of lifting, or lightness, or brightness, change of mood, etc, when applied regularly over the head. These folks weren't using it for depression, but some other ailment, so it was an interesting effect to hear about.
>
> However, before the above experience, I had my own experience. I was trying the magnetic pulser on my head for a vanity reason (gray hair), and after a few days I notice I had sort of a spring to my step and felt energetic and generally happier than usual. I thought to myself, my sleep is the same, diet is the same, what has changed? And then I realized the only thing different at that time was me using the magnetic pulser once a day for about 10-20 minutes over my head.....very interesting...
>
> The point of my story of course, is...$500 a treatment? The Neuro Star company can get stuffed!
>
> And yes, the Brain Tuner of course works the same way for most people.

>
> Warm regards,
>
> Vicki

Re: The Body Electric and more
5/6/13
Hi Mike:

There is more than you are ready to accept, so far, as I read your question... and Dr. Jenson is living in the past. Below I expand on what I have said already, really, in my (this) book: My book includes a good deal on this very topic. In fact, I actually included spirituality in ultimate (AHP) solution and I believe that it is a truly necessary part of healing and wellness. Furthermore, it goes on, saying that dreams and contemplation are necessary as a means of staying well. However, spirituality certainly is not the primary topic of my book. So, no. This is not about nutrients, Mike, unless you consider Love a nutrient... certainly, we all must have love to Thrive and spirituality is definitely about Love. In fact, it has been proven with babies... they simply die without love.

Do I have a reference, Mike? Not right off the top of my head, No. But I'll bet that you will accept the above as fact. While I often use the word Quantum and Spirituality in the same context, they are actually very different... well maybe not so much and here is why:

Quantum energy is defined as the vast energy present in the vacuum of space and at the zero point temperature (absolute zero), when there is no molecular movement. However, it turns out that many of our concepts of spirituality coincide closely with Quantum Physics (notice that Mike uses the terms interchangeably ;0). If spirituality is real, then the fact that the two virtually coincide in certain ways makes way too much sense to my way of thinking. In fact, they may prove to actually be the very same thing, once it is all sorted out. That would mean that the ancients were much more wise than we are inclined to give them credit for, as Quantum Machines, and we are actually relearning from the past. The question then becomes... how much have we forgotten?

If we are the Quantum Machines that my (this) book claims, then we have infinite access to Zero Point Energy and we can and do influence its actions. Astrophysicist Dan Nelson, my new mentor, makes this clear in his videos. These results have been confused by Newtonian Physicists and they are lumped into those areas separate from their beloved "science" that some call "hard" science. We now know that hard (Newtonian) science is badly flawed and that this fact is showing itself throughout the medical world, today, as well as in physics where there are no explanations that can cover observed Quantum Phenomena. Dan Nelson says that there is no such thing as metaphysics, a term often employed by hard scientists to mean "unreal" physics. So it is either real or not. And often, we are discovering that large parts of metaphysics are real.

So, I suggest that we have missed the boat as we have moved away from our inner knowledge and many still only accept hard science as proof. For example, the more that we have divorced love from medicine, the worse the results have become. The door to door doctor of the '50's had a great deal to offer and they learned through personal, first-hand experience (another part of spirituality) and not by watching a computer screen... Awareness IS love and first-hand experiences teach that. Those doctors with bags listened (another part of spirituality), paid close attention (still another), and learned. In other words, they gained inner wisdom. Watching a computer screen may inform you well technically, but it is not conducive with real wisdom. Interaction with others is really the key here. So, I suggest that we have lost the key.

Furthermore, when a doctor attempts to convince me that an artificial tooth, knee or hip is actually better than a real one, I know immediately that I am talking to someone who has so endorsed hard science that he has completely lost touch with reality. My opinion is that he is the one under an illusion, not me (as he wants me to believe), and it is time to move on. In fact, this has happened to me twice in the last few years. Since I generally have had few health issues, I am not encouraged by this. Each of those times, I have searched out and found my own answers. Fortunately, they were not life-threatening issues, so I have had enough time to find real solutions as I move on.

Regards, HUuman
>
> >Pam wrote,
in fact, spiritual wellness is probably, by far, the greatest part of wellness and health. <

> >The problem with this statement Pam is that when one begins to use words spiritual. experiences, quantum it opens a Pandoras box of the words meaning different things to different members. Your definition may be different than mine, and mine different then another's.<
>
> >So who is right? Is it you, Jim or myself? Then one begins to be bogged down such as I've found myself. Could you and Jim mean that by taking certain nutrients, brain performance is improved? Or that by taking certain nutrients your stress level is reduced, so that it is easier to feel pleasure, peace, and positive emotions toward others? If that is why is the message, then I agree, but to use these above buzzwords and then reject scrutiny from others when these expressions of thought is used then that is where we part. Btw this is food for thought, Dr Jensen taught that whenever phosphorous is in excess, then these (spiritual, floating in the sky)

manifestations occur. Could it be that some members are eating too much seafood? (big smile)
> mike<

Thanks, "Huuman"

55. Greatest Testimony Ever

It has been great to see the msm stories coming in. We thought Jim especially would be interested in our experience so far with it.

My husband was diagnosed with stage 4 mesothelioma-a cancer caused by asbestos exposure (during his years as a motor mechanic),with a poor prognosis- just over 2 years ago. Chemotherapy was our only treatment option, and we felt some pressure from our Specialist to try it. All this time though we were searching for alternative options and came across high level msm protocol on cancer tutor. 3 rounds of chemo showed no change to his tumor, so funding was stopped and we were able to pursue an alternative regime.

My husband built up to 9 tablespoons a day (of MSM), and he has regained good health. Our regime now is msm, magnesium, astaxanthin, he has optimized his vitamin d levels with sun exposure. We also take kefir with ground flaxseed daily and an interesting array of vege juices. We are now down to around 2 teaspoons of msm a day, but happily increase this if we feel the need. We have numerous friends and family on msm now and all report impressive improvements in their health-too many to describe in detail here.

Jim, we wondered what you might think about this. About 8 years ago-before we knew about my husbands ticking time bomb- we were asked by a friend to make a Bob Beck zapper for her husband, who had been diagnosed with melanoma in the brain. Because we had some concerns about his ability to communicate any issues he had using the zapper, my husband decided to do a 1 month stint on it along with the high level vit c. He felt great after using it, and we wondered on reflection if we may have inadvertently halted his tumor back then.

Deanery, New Zealand
(Editors Note: Paul Dean was still fine last I inquired, Jim)

56. 4th Edition Posts

Improved Energy, Mental Function and Removing Skin Lesions 9/3/2016

Hi James!

Just finished your wonderful book and can't tell you how much I enjoyed it! I have had an incredible experience with MSM. I have had cysts and skin lesions on my body, particularly my face, for over 10 years from an infection. During the past year, I have had success with the serrapeptase enzyme in reducing them, but since using the HL MSM protocol two weeks ago, the progress has been very dramatic.

At 50, I feel like a completely new 20 year old man with energy and vitality to spare. I can testify that MSM definitely improves brain performance, as I trade commodity futures for a living, which requires split second decision making. After going on the MSM, my daily take has gone from an average $500 (on a good day) to over $1500 consistently! I think a lot of this success has to do with MSM's reducing the anxiety I was experiencing while in trades

Thanks so much for writing your book and making this knowledge available to the public. I have spread the word about to a lot of people already, and am looking forward to doing a lot more!

Best Shep

References/Books on DMSO
June 2016

A good place to start is
DMSO Nature's Healer by Morton Walker.... often free on internet

Of special interest beyond books are patents by Jacob and Herschler...As you may know, Dr Robert Herschler "found" dmso as a byproduct of the NW timber industry and called it to the attention of Dr Stanley Jacob at u of Oregon. Lots of fascinating conclusions.

One might also consider reading everything written by
Janet/aka Garnet, list owner of DimethylSulfoxide-DMSO@yahoogroups.com
She has much experience, a great background in bio chem and knows more of DMSO than most laypeople. She has her own personal struggle at the moment so will not be available but if you read only her posts you will be very well educated re DMSO.
Because of the maturity of the list, not much is exchanged re DMSO, occasionally some of us older ones jump in to quell the paranoia about

purity and absorbing "everything" past the blood/brain barrier.

I personally know someone:
* whose scleroderma was, dare I say it?, cured using DMSO. She flew from FL to OR every 6 months and brought back several gallons on the plane.
* for whom it stopped a stroke; no damage whatsoever
* used it well for eyes, bruises, broken bones, sprains, burns, pains, etc

Several tidbits have served me well when recommending DMSO...

* DMSO only carries substances with a molecular weight of less than 1000 daltons. ....or at least smaller than it's own mw.
* A 2-5% solution is used in transporting transplant organs
* To those who would make purity claims of their purchases - If the DMSO you purchase has no odor, it likely is mostly water; it is immensely hygrophyllic.

Additional filtration does help Jim Ahart has
a section on his horseracing page that deals with his 2 micron filtering. (he also gets excellent herbal tinctures using DMSO
* DMSO is Not a strong solvent. It will not eat away at the plastic bottle it's shipped in---even if you think it's better because your supplier re-pours into glass bottles ;-)
* if you're only taking a few drops at a time, it is unlikely you will stink. the "cleaner" your system is, the less you stink--so they say.
* If it's useful on a million dollar animal, I'm not real concerned about danger to my body in appropriate volume/use
* Most research showed that it is most effective, without a histamine reaction, between 30 and 70%. That doesn't mean those with tough skin can't use it at 100%, it's just that you don't have to get the full effect.

Saralou

Our Greatest Threat is in the Name of Science (4th Edition)

The government hires scientists to support its policies
http://www.amazon.com/gp/product/1626360715?*Version*=1&*entries*=0

Industry hires them to support its business (without regard for the science)

Universities hire them to bring in grants that are handed out to support government policies and industry practices (yes, it is circular).

Organizations dealing with scientific integrity are designed only to weed out those who commit fraud behind the backs of the institutions where they work.

The greatest threat of all is the purposeful corruption of the scientific enterprise by the institutions themselves. The science they create is often only an illusion (I call it a religion), designed to deceive; and the scientists they destroy to protect that illusion are often our best. The book referenced above is about both, beginning with Dr. Lewis's experience, and ending with the story of Dr. Andrew Wakefield.

Finally, all science is selective in its derivation. In the past, much of what fell into the realm of science is not even considered today. The science of Remote Viewing http://www.greaterreality.com/rv/instruct.htmthat came under Reagan's Star Wars is now considered bunk. If it is not cold, hard objectivity per http://skepdic.com/remotevw.html it is a fraud and summarily dismissed. There are still plenty of older Russian scientists today from the USSR days who are dismissed as quacks by the US (and thus by Europe).

Science should be driven by what works, not what a few "believe in." When it is driven by beliefs, it is a religion. Today, in the US, it is a religion.

Medicine and drugs are a central part of this religion. If you ask doctors today what coconut oil does for you, they will tell you that all saturated fats will eventually kill you and this is one of them. The origin? Universities. Next they will offer to offset these effects by offering statin drugs, No disease escapes their ignorance.

Listen to the ads carefully. If you have arthritis, they have a drug that can kill you (and they admit it freely), but it will relieve joint pain to some degree (not make you well). WHAT? Why does our government allow such ignorance yet not allow someone selling an organic orange peel for health without years of study?

The Water Device (4th Edition)
Mon Jun 13, 2016 5:18 am (PDT) . Posted by:
"Jim Clark" huuman60
*Now adding to the HL MSM and water stories which both can a part a

huge part in cancer incidence:

http://articles.mercola.com/sites/articles/archive/2016/06/11/nutrition-influences-cancer.aspx?utm_source=dnl&utm_medium=email&utm_content=art1&utm_campaign=20160611Z1&et_cid=DM107622&et_rid=1525390246

If you have been reading my posts and books over the years, you know that my friend, Dr. Richard Price, and I have focused on mitochondria. Especially, mitochondrial replication (that is, raising the average number of mitochondria per cell) so as to increase energy levels and also to raise disease resistance and as the talk above suggests.. So cancer resistance. This is all accomplished with Fractals and our "Living Water," as Rich calls it.

My average mitochondria count per cell is still holding at 2000 per cell. Babies are lucky to be born with 800mpc. Rich put me on that particular fractal stack a year ago designed to do this. When he does this, your average cellular age drops dramatically. I have discovered that there are other ways of doing it, but it is not not uncommon for a 35 year old to have a cellular age of an 8 year old. This does not make you look like a child, but it raises your energy levels and autoimmunity equal to that of an eight year old. Mine is hold at that of a 12 year old. So I do not tire when playing tennis.

This lecture by Dr. Gary Fettke above basically agrees with my last two books, "Methylation, Awareness and You," and "Beyond Epigenetics." So it is good that we are now hearing from mainstream scientists and, in this case a surgeon. His emphasis on the Ketogenic diet and nutrition, in general, and in particular, his suggestion that spirituality and attitude are directly related to healing aligns with all three of my first three books and is very topic of my fourth book, "Surrender... to the Oneness."

So, as we move deeper into this 21st Century and, hopefully, more and more divorce ourselves from these crazy methods of especially the last sixty years that as Dr. Fettke points out have not only been ineffective in cancer, but have raised the incidence of it continually, lets hope that this attitude catches on. The results can change everything and quicker than even he imagines.

Electrical Healing(4th Edition)
Tuesday 2/28/ 17 2:12 pm (PDT) . Posted by:
"Jim Clark" huuman60

I have one root canal (a popcorn event). And, yes, root canals harbor many infection problems as they are no longer biologically a part of you. But dentures pose a different threat and no one wants migrating teeth. Dentures too should be avoided as well as any metal in your mouth. Metal is an electrical vibration. Still, I ate the popcorn and fractured the tooth... a physical fact.

But herein, I offer a remedy for fellow popcorn lovers: Whenever I have a gum flair-up or other mouth infection resulting from the captured bacteria, a small amount of DMSO on my finger rubbed on the problem area relieves it in minutes. It is uncanny.

The bottom line is that I have a good, fast, and cheap answer. Therefore, were I to snap another tooth, I would still opt for keeping the dead tooth and getting a crown with a root canal (which is contrary to even the most wise of healers who suggest no foreign structures in your mouth). Taking this a bit further, at least nature made my foreign object (the dead tooth) and it was not approximated by a dentist, who is guessing to some degree. .

We do not discuss Bob Beck much anymore in the Beck Groups and have not for years, really. Of course, his electronic healing still works. Just because he died and has done no new work does not negate his brilliant work. We have said about as much as you can about what he did on the web. The man was a genius and his discoveries absolutely work. As it were, I happened onto MSM/DMSO about the same time as I came across the Beck Protocols, watched his lectures and read his work, and used his devices. Before his fatal fall, Bob had developed his system and had proven it out to himself and many others.

Obviously, all that Bob Beck did conflicted with the mainstream knowledge even today for the most part, but when you really study it and do your homework, there is plenty of good work out there to scientifically back his discoveries. One of the best of those sources is, "The Body Electric," by Robert O. Becker, MD and Gary Selden. For those interested, this book and Becker's work is the gold standard IMHO. Before you venture into electronic healing, you should read Becker's book and certainly, you should watch all of Bob Beck's lectures on YouTube (as long as Google allows them to exist, which could end shortly as we see the alternative fields being destroyed by them as we speak). Obviously, big pharma has money and pays Google's bills even if it is almost all total illusion.

In his book, Becker apologized for being, "just an MD" and not a pure scientist. That is nearly comical today as surgeons and other unrelated specialist MD's, who have no clue about diet or body chemistry, go on the web and mainstream news as "experts" with "their discoveries" ranging from alpha lipoic acid to xanthin oxidase expressing their wealth of knowledge in these areas. Most of the mainstream buys their pathetically misunderstood knowledge. In fact, few if any of them ever did any original work to find their discoveries and, yes, these all have some application in healing. Realize that the typical doctor gets one course in nutrition and mainstream nutrition overall today is almost totally clueless. In fact, being a fully trained mainstream nutritionist is a detriment for the most part. Therefore, you seldom get good nutritional information from the mainstream media. I always listen though, just to get a good laugh.

Therefore, keep in mind that almost no MD, as mainly a pill pusher for pharmaceuticals, has a vast study knowledge in alternative health measures such as vitamins, minerals, antiaging, and wellness. Also, no amount of surgeries will get a doctor any closer to true knowledge of this outrageously useful alternative field that is making new progress daily. Finally, knowing a disease's name, and having a true diagnosis of it is only a first step in effecting a cure. In fact, a true healer only needs an understanding of the vibratory rates within an individual to effect a cure, period . Furthermore, even if the patient did not reveal the symptoms of their disorder, it is always about system imbalances and not a war against offending germs as commonly agreed within the mainstream professions. The bottom line: Balance the systems and cures occur from within. We always cure ourselves. No cure ever comes from an outside source, thus, they are truly spiritual events.

Therefore, Becker may not have totally understood what he saw, but he was certainly much closer than the common pharmaceutical "cures" of today. Frequencies explain it all and, yes, electricity itself can create balances. Still, these methods are really hit or miss to some degree. Therefore, Bob Beck's devices do indeed work and often, but they also can easily miss. Yet, in the hands of a tuned-in healer, they can be extremely effective and are often useful.

Now, let's go back to why the DMSO cures the above gum infection and this is key to all of this. MSM and, the active oxidant reciprocal, DMSO affect the Methylation pathways, and directly.

DMSO, itself, is probably the most direct and efficient healer on earth that can be taken externally or internally. However, that said, a good vibratory healer can duplicate and even exceed DMSO's effects using fractals and charged water among other devices as my friend, Dr. Richard Price does. DMSO and MSM, especially when used the with powerful nutritional supplements that the above "surgeons" sometimes recommend (with no real understanding of what their cures are doing), literally lights up the mitochondria, supercharging them, and thus making each individual cell in your body an energy factory that heals.

Over time, high levels of MSM, taken daily, helps maintain the above levels. My book, "It's the Liver Stupid" discusses how MSM, when taken orally literally transforms your cells, picking up an oxygen molecule, thus becoming DMSO and a healer.

These are the "Oxygen Transport Pair" and they work together to heal when one or the other is ingested or DMSO is applied topically. The mystery of why people were being healed by this pair is over. Now I know the mechanism for the reported outrageous heart disease and cancer cures that made no sense. Not only does this work for, so called "dis-ease," it works to help you avoid physical traumas such as fractures and joint displacements (in overall wellness, not just healing) that are commonly attributed to normal aging. Also, if the stress is great enough, and emergency surgery is the only answer, they speed the rate up of the natural healing process and this is all still vibrational in its effect.

This is your ball game. You have control of your life. Take the above information and apply it in ways that suite you as an individual. Take the above with a grain of sea salt, but gather your own salt and make your own mixtures. Pay attention, enjoy the game and help extend it to extra innings... but easy on the popcorn

Antiaging Indicates Health (4th Edition)

First off, it is all about your diet and liver (microcellular health). Most people eat what their parents ate. Unlike what we are told, these bad habits are passed on, mostly, not by genes. High carbs: grains, sweets, breads and few green foods and likely, your hair will prematurely gray.

Now, switch to the LCD, HL MSM, undenatured whey, and, if you can afford them, my new star, Liposomal enzymes. Mysteriously, your hair will darken. Duh!

A reversal in aging is a far better indication that all of what the doctors measured in the studies below and these aging factors will not affect you ever again if you stay with these new habits that caused the obvious reversals.

Second, the inner physical changes associated with aging invite impending major illnesses like cancer and heart disease. The outer signs of recovery are visible to everyone, such as: skin wrinkles begin disappearing, as the voice, ears, and eyes come back. Quickly, finger nails, toe nails, and hair become fast growing. Prostrate problems disappear and even beard growth is accelerated in men as further indications. Hands shake decreases as and balance improves.

These were all age related signs and thus previewed death. We simply do not need blood pressure monitoring or expensive tests to see these come on as the bad indicators disappear as we antiage.

Most people, while they are interested in these things, could most likely benefit from professional advice from someone who is aware of the more esoteric things as liposomes, which have been slowly accepted into the pharmaceutical culture since 1965 for drug applications. These are extremely dangerous when applied to drugs, since they bypass the body's natural defenses. However, when incorporated into nutrients they can get them into your body without having to go through the energy intensive digestive process. So they are a God-send for supplements, if you can afford them. Expect many more varieties of them to be available in both drugs and nutrients in the coming years.

Unlike Liposomes, few experts know anything about the effects of MSM and DMSO in high levels. You are basically on your own here and can likely teach the experts a thing or two if you understand them. They are as simple as liposomes are complex.

Kind Regards,

Jim (Huuman)

Liposomes for Health (4th Edition)

https://www.deepdyve.com/lp/elsevier/catalase-conjugated-liposomes-encapsulating-glucose-oxidase-for-LrJ5anpXKE

These Liposomals are the most intelligent and ground-breaking method of mechanically getting to disease that we currently have. They are the alternative medicine equivalent to what mainstream drugs and antibiotics have promised, but have fallen short on.

Dr. Wallach told us twenty years ago that colloidal minerals were the way to health and he was correct. These take colloids one step further and pack them into a phospholipid bilayer membrane with an aqueous drop that the body immediately recognizes as food using electromagnetically activated hydrogen peroxide. The actual encapualization can be done in several ways as outlined in the above article.

They, unlike most foods, are unaffected by high temperatures and can be stored for long periods.

The result is essentially an ideal supplement that is highly bioavailable and due to the small molecular size can be utilized transdermally is many cases as well as orally.

Below are some of the incredible products now available. Read the claims. These are not like the ridiculous claims that you commonly hear about drugs on TV. They actually do what they claim.

https://www.deepdyve.com/lp/elsevier/catalase-conjugated-liposomes-encapsulating-glucose-oxidase-for-LrJ5anpXKE These Radical Scavenging Enzymes (best taken transdermally when possible) are the most powerful free radical scavengers currently known.

BENEFITS

- ANTIAGING PREVENTATIVE
- SKINCARE
- WOUND & BURN CARE
- INTERNAL ANTIOXIDANT SUPPLEMENTATION
- HAIR COLOR RESTORATION & PROTECTANT
- USE DURING OXIDATIVE MEDICINE THERAPY TO PREVENT OVER-OXIDATION

And Again, these are not idol claims, and the basis starts with mitochondrial revitalization through Superoxide Dimutmutase and Catalase (SOD). The mechanics are not new, we just never had the method to make this occur. Again, first cost is high, but these should last you a very long time. This is just one of several for sale.

If you have a health problem or you want to maintain your wellness and youth, this is a direct path.

Liposomal C:

As ascorbic acid: https://treato.com/Vitamin+C/?a=s#drugFacts Those of us who have been around have mostly discovered that Linus Pauling was wrong with very few exceptions. However, Liposomal C makes him dead accurate.

This, when administered IV will, as reported by some adventurous doctors, cure cancer. It could be considered like eating a half- bushel of lemons, skins and all every day without the obvious downside. No disease can hold up to this firepower. Its like blowing up an ant with a **stick of dynamite. Add the above enzymes and who knows? Would it become the Mother of all Bombs?**

https://www.nanonutrausa.com/product/liposomal-vitamin-c/?utm_source=bing&utm_medium=cpc&utm_campaign=Vitamin%20C%20-%20Search%20-%20USA&utm_term=%2Bvitamin%20%2Bc&utm_content=Generic%20Vitamin%20C

Hey, I could list ten others and they are all good. Read the effects and shop. I have given you ways to make your own that are not as effective, but are very good. These were the first liposomal source. Watch because virtually all supplements can be made by this process. If it was good before, expect it to be great now. Again, these claims actually make sense, unlike drugs.

Some others supplements to consider (and there are plenty more):

Lipo Turmeric: https://www.amazon.com/s/?ie=UTF8&keywords=liposomal+turmeric&tag=mh0b-20&index=aps&hvadid=7015127679&hvqmt=e&hvbmt=be&hvdev=c&ref=pd_sl_89qsw38nk_e

Lipo Magnesium: https://www.amazon.com/Liposomal-bis-glycinate-Formulated-Seeking-Health/dp/B00BJO3YJ0/ref=sr_1_1_a_it?ie=UTF8&qid=1492272649&sr=8-1&keywords=liposomal+magnesium

Colloidal minerals (alone) could be close enough to make the case that this is not the best way to spend your money. But you can just spray this on after a shower

Lipo B-12 Methylcobalamin https://www.amazon.com/Methylcobalamin-Phosphatidylcholine-Phosphatidylethanolamine-Seeking-Health/dp/B00F0T41GU/ref=pd_sim_121_3?_encoding=UTF8&pd_rd_i=B00F0T41GU&pd_rd_r=NAX045T9YAJASMGE5D9P&pd_rd_w=ERTde&pd_rd_wg=h2llc&psc=1&refRID=NAX045T9YAJASMGE5D9P If the reason you are not well is that you lack this important B vitamin, you are going to be well soon, but it costs $2 a day to take. What price wellness? Maybe, if you are well, you supplement with organic B-12 normally and take this once a week or if you start feeling badly, you jump in, keeping it around.

Liposomal B Complex Plus http://try.livonlabs.com/camp05135/?prod=liposomal&gclid=CMD-75jwptMCFZGKswodX4ADVw A way to take all B's

Lipo Astaxanthin Trial Completed https://clinicaltrials.gov/ct2/show/NCT02397811

Lipo Astaxanthin should be high on anyone's list when it is available and Cyanotech will likely be the first to sell it.

Liposomal GSH (4th Edition)

I think it was a simple question, how is everyone measuring Glutathione Serum in blood levels? As you yourself have pointed out, GSH liposomals are not cheap...

Why did you pick Liposomal GSH out of all of the Liposomal possibilities when we already know that these are not commonly bioavailable otherwise as supplements? Also, MSM, Undenatured Whey, and the LCD are simple to implement and they work.

Truth is that, just as Duncan Crowe told us ten years ago, and it is still true, whey is the way to raise your glutathione levels. Of all of these, GSH liposomals may simply not do anything or they may cause problems as N-Acetyl-L-Cystine (NAC). GSH could possibly throw your body out of balance. I would avoid it till we know more unless you are interested in being a guinea pig.

Sam-E, not cheap either, may be an optional way to glutathione, somewhat, through the back door, but MSM, Whey, and the LCD are certainly more direct. What Dunc was saying is that your body knows when to produce glutathione. Give your body a chance to regulate itself by supplying it with the correct raw materials. It knows the vibrations that you need to balance. Your mind does not.

The other Liposomal supplements are most likely winners even though Liposomal GSH may not prove out long term. Even if Liposomal GSH does work, you are not going to prove it out short-term, especially if you are already following these far less expensive interventions.

Kind Regards,
Jim (huuman)

Mike Adams and Politics (4th Edition)

http://www.naturalnews.com/2017-03-22-the-big-lie-upon-which-all-government-is-founded.html

My friend, Mike Adams, the health Ranger, who owns and operates one of the world's most sophisticated food and chemical testing labs, speaks out about the big lie... that any politician can truly represent your interest at heart. Certainly, any group of them can't come close to representing your particular needs as they all claim to be able to do.

Today, we have a two party system in a very large and diverse country that would serve us better were it ten countries or more. This huge system does not have my interests in mind at all today and I suggest that it was already at its limit in 1776. For example, just today, they will argue how they should take my stolen IRS money, a recent invention, and fund all of those who rely on the subversively run US drug market previously called Obamacare.

Since I have learned, using my own creativity (you may call it luck, but plenty of others who have followed my protocols who exhibit that luck), I have zero need for this ridiculous approach to health and wellness, But not one "representative" out there has a slight clue as to what we do or why what we do works for us. Mike Adams knows a lot, but even he has plenty to learn here. Health, several hundred years of wellness, and antiaging are your birthright.

Instead, these idiots push such crazy ideas as "Statin: Drugs that actually cause the "diseases" that they were designed to stop, "heart disease and strokes." These, very effectively, in doing what they were designed to do, decrease serum cholesterol levels. in doing that, they literally help the users "catch" Cancer. If even a few of these "representatives" actually had a clue as to what has been known in the most advanced practitioner's community for the last ten years: That we really thrive on saturated fats... something that my cats intuitively know (they always eat the pure fat first)... they would abandon the drug company supplementation totally and the health care market generally. I say, supplementation, because there are millions of people out there under the false assumption that they should supplement their diets with any number of truly deadly drugs, like statins and aspirin, for the rest of their lives (which they are unknowingly decreasing).

You may say, but this is the exception, not rule. Actually, this is absolutely the rule. Please listen carefully to the drug community's ads. In general, there is no safe drug, but that is another discussion. This outlay, though, comprises a good half of our tax money taken as a whole that our representatives are striving to give us greater access to.

Don't get me wrong here, there is a need for some enforced regulation of our actions, and, with that, planning, and controlled leadership.

Jim

The Vibrations of Life Feed You (4th Edition)

https://www.freemansperspective.com/last-time-held-baby/

Hold a baby... and the reason is, more than expressed above, that they know more spiritually than you likely do, or you are less likely to recall anytime soon, if you do not recall why. You, with few exceptions, have forgotten who you are. Listen to kids, but most have forgotten the truth after five years old and many before four. We all become products of mind which is in a constant war against the love that you once held.

This is mainly about human babies, but they are around much longer. Any new born baby holds what I discuss and we maintain what we are given a few years after gaining our communication skills which adds in the lessons.

Our so-called, modern world is fairly devoid of the truth. Science, our new god, teaches the opposite. it relies on overwhelming fact in ways that keep you out of touch with the true reality, God or Love. Holding a baby will help you get that Love back (and it is absolutely contagious). You absorb Love through your skin in the form of frequencies that you long ago disassociated yourself from. More on that below:

Don't get me wrong here, our best scientists are those who maintain what I am discussing here. It encourages creativity, which is the essence of discovery. This, in turn, melts the invisible barriers that screen you from the truth. You simply stop keeping those barriers and become an independent absorber of truth.

When it comes to accepting or rejecting what we know, there are two simple ways of doing this. One is to die and start over, something that Christians have been taught is not possible. However, if they read John in the NT, it clearly happens. The other is to survive for many years as did Methuselah in the OT. The value of a long survival is that you become extremely wise over time... something that is not even known today by the Main Stream Media and our spiritually challenged science, but certainly it occurs with the Spiritual Masters that live among us and keep things balanced. This all sounds crazy to most everyone today and defies our media reporting. We are taught to die at 85 (or so) years and our entire social structure relies on that occurring. Challenge that structure and expect blow back. One thing for sure, if you do not grow spiritually along with the tricks that I hand out freely here in "It's the Liver Stupid," you will die at the young age just as we are taught and start over.

Read this again if you already did. The above just adds to the story. The idea is to get back what you lost from the teachings of society. You must stop questioning the simple things that an open heart reveals. No, you do not have to start over to even have a chance. Also. how many get it when they do? It is almost that simple, but oh so hard to grasp. In a sense, if you lost what you had in the first five years, babies are a 2nd chance to get what you have forgotten, just as the spiritual masters are our 3rd who keep and maintain it.

When you start over from birth, your spiritual memory is generally wiped within the first five years of your new life. Children are born, generally, with nearly a full memory of their last life and even some of the higher worlds. They will say things like, "When I was boy, I did so and so..." and we will laugh (teaching them to forget).

Children, by in large, are the most intelligent people on earth, spiritually. This helps make them the most loved and protected people in society. When you finally are allowed access to a spiritual master, and there are several hundred who work here at least on occasion, you will begin to feel the sound and light (or frequencies and love) that they radiate. This is pure, unattached love.

Some masters keep a rather "permanent" body here (by our ignorant standards). Others have important jobs elsewhere in this universe and beyond. The one that we refer here to as "God" (by our religions) is totally in charge of running Earth and It is a 13th initiate. God has authority here over some of the most profound of masters like Rebezar Tarz, a 32nd initiate, who maintains a permanent human body in the high mountains of Tibet. No master is here to make a splash spiritually as the Christians teach of Jesus. It is all about a low profile and keeping the balance. This physical world is and must be balanced between what we call good and evil, the karma of eastern religions. These are all illusions. The more we hold onto them, the more real they become to our minds, so the lesson here is to let go of the negatives as they crop up.

I have reported on a master before here who lives in Lome, Togo (West Africa). he is 665 years in this human body and he has helped me quite a bit. He even issued me one of his most gifted students, a recent and highly talented 5th, to help me in my work.

There are spiritual masters throughout earth and they can maintain several human bodies at once, when needed. In addition, if there is a need, a master can, with agreement, take over any body, human or otherwise, and when needed, be, will do just that. They are unlimited and can be the kindly lion that you see on TV that acts with love, or the pizza chef who treats everyone so lovingly on a given day. Your job, as a spiritual being is to recognize them, the love coming out, and to honor it. Simple! Your gift then, is to just feel that love. True spiritual, unattached, love is mind blowing and trans-formative.

There are many ways to extend your life. All of them include vibratory measures of one kind or another. One of them, reported by Tre, in the CCO group, is a vibratory mat. This is an expensive augmentation, but it is valid. Minerals and vitamins are nothing more than vibrations. Most man-made artificial vitamins are too harsh and are, thus destructive. You need not ingest anything to get its vibratory contribution. Sunlight, for example, is vibratory. Too much of anything vibratory will make you ill.

Few reading this are quite ready yet to sit in the presence of a master and doing that could make them ill. However, the higher you are spiritually the more you can handle as a rule. Some supplements, like astaxanthin allow more ingestion of sunlight, and I recommend them in my book. On the other hand, Sun block as recommended by the media and mainstream, stop them, so I suggest their disuse.

In fact, nearly every drug that is distributed today over-stimulates your frequencies and literally serves to kill you, long term. Even aspirin and pain killers can do this, long term. Pain is nature's way of warning you that your frequencies are out of balance and must be rebalanced. There are no "free lunches." You, as a spiritual being, must do your part. You were given a mind and the spiritual tools required to become a master in your own right. Even if you fight this process, resisting the tools, you must eventually accept the love and surrender to it, as my 4th book, "Surrender to the One" tells you. This spiritual process is in no way flawed and never does God desert you as some proclaim. However, no one said that this would be easy.

Kind Regards,
Huuman (Jim)

A Great Review (4th Edition)

Unbelievable (Amazon)
By California Sandyon August 22, 2015
Format: Kindle Edition|Verified Purchase
"I have had rheumatoid arthritis and osteoarthritis for 20+ years. I have never taken prescription drugs as recommended by my doctor and have had some success with a naturopathic doctor. The MSM protocol in this book has given me my life back. It has only been a couple of weeks and the only way anyone can tell I have arthritis at all (including me) is by the typical rheumatoid deformity in my hands, which is quite advanced. Interestingly,, my nearly chronic sore throat of the past 2 or 3 years is completely gone. I'm talking about arthritis in nearly every joint (even my toes), except my spine. I'm taking about half a cup per day. I can run and get down on my hands and knees again for the first time in years. I have also been able to discontinue about two or three hundred dollars worth of supplements per month. Thank you Jim. Amazing!!!"

I thank Sandy on for the great review, but wonder if her fingers ever straightened out? My aunt died of breast cancer before I knew that MSM/DMSO was even around. Had I known about it then, I may have fixed both problems at once for her. That would have been awesome. She was one of the most wonderful people that I have ever known
Please, if anyone here has a story like Sandyon's, email me and post a review. This is how we learn and why the information in this book is so very useful.

Kind Regard,
Huuman (Jim)

Stephanie Senneff (5th Edition)

This basically wraps up all that this book lays out in no uncertain term. Stephanie does not know a lot of what I say here in terms of supplementation, but she has a powerful handles on what is going on otherwise and there are no contradictions here:

by Stephanie Seneff [PhD]

seneff@csail.mit.edu
September 15, 2010

1. Introduction

Obesity is quickly becoming the number one health issue confronting America today, and has also risen to epidemic proportions worldwide. Its spread has been associated with the adoption of a Western-style diet. However, I believe that the widespread consumption of food imports produced by U.S. companies plays a crucial role in the rise in obesity worldwide. Specifically, these "fast foods" typically include heavily processed derivatives of corn, soybeans, and grains, grown on highly efficient mega-farms. Furthermore, I will argue in this essay that one of the core underlying causes of obesity may be sulfur deficiency.

Sulfur is the eighth most common element by mass in the human body, behind oxygen, carbon, hydrogen, nitrogen, calcium, phosphorus, and potassium. The two sulfur-containing amino acids, methionine and cysteine, play essential physiological roles throughout the body. However, sulfur has been consistently overlooked in addressing the issues of nutritional deficiencies. In fact, the American Food and Drug Administration has not even assigned a minimum daily requirement (MDR) for sulfur. One consequence of sulfur's limbo nutritional status is that it is omitted from the long list of supplements that are commonly artificially added to popular foods like cereal.

Sulfur is found in a large number of foods, and, as a consequence, it is assumed that almost any diet would meet the minimum daily requirements. Excellent sources are eggs, onions, garlic, and leafy dark green vegetables like kale and broccoli. Meats, nuts, and seafood also contain sulfur. Methionine, an essential amino acid, in that we are unable to synthesize it ourselves, is found mainly in egg whites and fish. A diet high in grains like bread and cereal is likely to be deficient in sulfur. Increasingly, whole foods such as corn and soybeans are disassembled into component parts with chemical names, and then reassembled into heavily processed foods. Sulfur is lost along the way, and there is a lack of awareness that this matters.

Experts have recently become aware that sulfur depletion in the soil creates a serious deficiency for plants [Jez2008], brought about in part by improved efficiency in farming and in part, ironically, by successful attempts to clean up air pollution.

....

7. Sulfur and Alzheimer's

With an aging population, Alzheimer's disease is on the rise, and it has been argued that the rate of increase is disproportionately high compared to the increase in the raw number of elderly people [Waldman2009]. Because of a conviction that the amyloid beta plaque that is a signature of Alzheimer's is also the cause, the pharmaceutical industry has spent hundreds of millions, if not billions, of dollars pursuing drugs that reduce the amount of plaque accumulating in the brain. Thus far, drug trials have been so disappointing that many are beginning to believe that amyloid beta is not the cause after all. Recent drug trials have shown not only no improvement, but actually a further decline in cognitive function, compared to placebo (New York Times

Article). I have argued elsewhere that amyloid beta may actually be protective against Alzheimer's, and that problems with glucose metabolism are the true culprit in the disease.

Once I began to suspect sulfur deficiency as a major factor in Americans' health, I looked into the relationship between sulfur deficiency and Alzheimer's. Imagine my surprise when I came upon a web page posted by Ronald Roth, which shows a plot of the levels of various minerals in the cells of a typical Alzheimer's patient relative to the normal level. Remarkably, sulfur is almost non-existent in the Alzheimer's patient's profile.
To quote directly from that site: "While some drugs or antibiotics may slow, or if it should happen, halt the progression of Alzheimer's disease, sulfur supplementation has the potential of not only preventing, but actually reversing the condition, provided it has not progressed to a stage where much damage has been done to the brain."

"One major reason for the increase in Alzheimer's disease over the past years has been the bad reputation eggs have been getting in respect to being a high source of cholesterol, despite the fact of dietary intake of cholesterol having little impact on serum cholesterol - which is now also finally acknowledged by mainstream medicine. In the meantime, a large percentage of the population lost out on an excellent source of sulfur and a host of other essential nutrients by following the nutritional misinformation spread on eggs. Of course, onions and garlic are another rich source of sulfur, but volume-wise, they cannot duplicate the amounts obtained from regularly consuming eggs."

Why should sulfur deficiency be so important for the brain? I suspect that the answer lies in the mysterious molecule alpha-synuclein, which shows up alongside amyloid-beta in the plaque, and is also present in the Lewy Bodies that are a signature of Parkinson's disease [Olivares2009]. The alpha-synuclein molecule contains four methionine residues, and all four of the sulfur molecules in the methionine residues are converted to sulfoxides in the presence of oxidizing agents such as hydrogen peroxide [Glaser2005]. Just as in the muscle cells, insulin would cause the mitochondria of neurons to release hydrogen peroxide, which would then allow the alpha-synuclein to take up oxygen, in a way that is very reminiscent of what myoglobin can do in muscle cells. The lack of sufficient sulfur should directly impact the neuron's ability to safely carry oxygen, again paralleling the situation in muscle cells. This would mean that other proteins and fats in the neuron would suffer from oxidative damage, leading ultimately to the neuron's destruction.

In my essay on Alzheimer's, I argued that biologically pro-active restriction in glucose metabolism in the brain (a so-called type-III diabetes and a precursor to Alzheimer's disease) is triggered by a deficiency in cholesterol in the neuron cell membrane. Again, as in muscle cells, glucose entry depends upon cholesterol-rich lipid rafts, and, when the cell is deficient in cholesterol, the brain goes into a mode of metabolism that prefers other nutrients besides glucose.

I suspect that a deficiency in cholesterol would come about if there is insufficient cholesterol sulfate, because cholesterol sulfate likely plays an important role in seeding lipid rafts, while concurrently enriching the cell wall in cholesterol. The cell also develops an insensitivity to insulin, and, as a consequence, anaerobic metabolism becomes favored over aerobic metabolism, reducing the chances for alpha-synuclein to become oxidized. Oxidation actually protects alpha-synuclein from fibrillation, a necessary structural change for the accumulation of Lewy bodies in Parkinson's disease (and likely also Alzheimer's plaque) [Glaser2005]
8. Is The Skin a Solar-Powered Battery for the Heart?

The evidence is quite compelling that sunny places afford protection from heart disease. A study described in [Grimes1996] provides an in depth anaylsis of data from around the world showing an inverse relationship between heart disease rates and sunny climate/low latitude. For instance, the cardiovascular-related death rate for men between the ages of 55 and 64 was 761 per 100,000 men in Belfast, Northern Ireland, but only 175 in Toulouse, France. While the obvious biological factor that would be impacted by sunlight is vitamin D, studies conducted specifically on vitamin D status have been inconclusive, with some even showing a significant increased risk for heart disease with increased intake of vitamin D2 supplements [Drolet2003].

I believe, first of all, that the distinction between vitamin D3 and vitamin D3-sulfate really matters, and also that the distinction between vitamin D2 and vitamin D3 really matters. Vitamin D2 is the plant form of the vitamin -- it works similarly to D3 with respect to calcium transport, but it cannot be sulfated. Furthermore, apparently the body is unable to produce vitamin D3 sulfate directly from unsulfated vitamin D3 [Lakdawala1977] (which implies that it produces vitamin D3 sulfate directly from cholesterol sulfate). I am not aware of any other food source besides raw milk that contains vitamin D3 in the sulfated form. So, when studies monitor either vitamin D supplements or vitamin D serum levels, they're not getting at the crucial aspect for heart protection, which I think is the serum level of vitamin D3 sulfate.

Furthermore, I believe it is extremely likely that vitamin D3 sulfate is not the only thing that's impacted by greater sun exposure, and maybe not even the most important thing. Given that cholesterol sulfate and vitamin D3 sulfate are very similar in molecular structure, I would imagine that both molecules are produced the same way. And since vitamin D3-sulfate synthesis requires sun exposure, I suspect that cholesterol sulfate synthesis may also exploit the sun's radiation energy.

Both cholesterol and sulfur afford protection in the skin from radiation damage to the cell's DNA, the kind of damage that can lead to skin cancer. Cholesterol and sulfur become oxidized upon exposure to the high frequency rays in sunlight, thus acting as antioxidants to "take the heat," so to speak. Oxidation of cholesterol is the first step in the process by which cholesterol transforms itself into vitamin D3. Sulfur dioxide in the air is converted nonenzymatically to the sulfate ion upon sun exposure. This is the process that produces acid rain. The oxidation of sulfide (S-2) to sulfate (SO4-2), a strongly endothermic reaction [Hockin2003], converts the sun's energy into chemical energy contained in the sulfur-oxygen bonds, while simultaneously picking up four oxygen molecules. Attaching the sulfate ion to cholesterol or vitamin D3 is an ingenious step, because it makes these molecules water-soluble and therefore easily transportable through the blood stream.
Hydrogen sulfide (H2S) is consistently found in the blood stream in small amounts. As a gas, it can diffuse into the air from capillaries close to the skin's surface. So it is conceivable that we rely on bacteria in the skin to convert sulfide to sulfate. It would not be the first time that humans have struck up a symbiotic relationship with bacteria. If this is true, then washing the skin with antibiotic soap is a bad idea. Phototrophic bacteria, such as Chlorobium tepidum, that can convert H2S to H2SO4 exist in nature [Zerkle2009, Wahlund1991], for example in sulfur hot springs in Yellowstone Park. These highly specialized bacteria can convert the light energy from the sun into chemical energy in the sulfate ion.

Another possibility is that we have specialized cells in the skin, possibly the keratinocytes, that are able to exploit sunlight to convert sulfide to sulfate, using a similar phototrophic mechanism to C. tepidum. This seems quite plausible, especially considering that both human keratinocytes and C. tepidum can synthesize an interesting UV-B absorbing cofactor, tetrahydrobioptin. This cofactor is found universally in mammalian cells, and one of its roles is to regulate the synthesis of melanin [Schallreut94], the skin pigment that is associated with a tan and protects the skin from damage by UV-light exposure [Costin2007]. However, tetrahydrobiopsin is very rare in the bacterial kingdom, and C. tepidum is one of the very few bacteria that can synthesize it [Cho99].

Let me summarize at this point where I'm on solid ground and where I'm speculating. It is undisputed that the skin synthesizes cholesterol sulfate in large amounts, and it has been suggested that the skin is the major supplier of cholesterol sulfate to the blood stream [Strott2003]. The skin also synthesizes vitamin D3 sulfate, upon exposure to sunlight. Vitamin D3 is synthesized from cholesterol, with oxysterols (created from sun exposure) as an intermediate step (oxysterols are forms of cholesterol with hydroxyl groups attached at various places in the carbon chain). The body can't synthesize vitamin D3 sulfate from vitamin D3 [Lakdawala1977] so it must be that sulfation happens first, producing cholesterol sulfate or hydroxy-cholesterol sulfate, which is then optionally converted to vitamin D3 sulfate or shipped out "as is."

Another highly significant feature of skin cells is that the skin stores sulfate ions attached to molecules that are universally present in the intracellular matrix, such as heparan sulfate, chondroitin sulfate, and keratin sulfate [Milstone1994]. Furthermore, it has been shown that exposure of the melanin producing cells (melanocytes) to molecules containing reduced sulfur (-2) leads to suppression of melanin synthesis [Chu2009], whereas exposure to molecules like chondroitin sulfate that contain oxidized sulfur (+6) leads to enhancement of melanin synthesis [Katz1976]. Melanin is a potent UV-light absorber, and it would compete with reduced sulfur for the opportunity to become oxidized. It is therefore logical that, when sulfur is reduced, melanin synthesis should be suppressed, so that sulfur can absorb the solar energy and convert it to very useful chemical bonds in the sulfate ion.

The sulfate would eventually be converted back to sulfide by a muscle cell in the heart or a skeletal muscle (simultaneously recovering the energy to fuel the cell and unlocking the oxygen to support aerobic metabolism of glucose), and the cycle would continually repeat.

Why am I spending so much time talking about all of this? Well, if I'm right, then the skin can be viewed as a solar-powered battery for the heart, and that is a remarkable concept. The energy in sunlight is converted into chemical energy in the oxygen-sulfur bonds, and then transported through the blood vessels to the heart and skeletal muscles. The cholesterol sulfate and vitamin D3 sulfate are carriers that deliver the energy (and the oxygen) "door-to-door" to the individual heart and skeletal muscle cells.

Today's lifestyle, especially in America, severely stresses this system. First of all, most Americans believe that any food containing cholesterol is unhealthy, so the diet is extremely low in cholesterol. Eggs are an excellent source of sulfur, but because of their high cholesterol content we have been advised to eat them sparingly. Secondly, as I discussed previously, natural food plant sources of sulfur are likely to be deficient due to sulfur depletion in the soil. Thirdly, water softeners remove sulfur from our water supply, which would otherwise be a good source. Fourthly, we have been discouraged from eating too much red meat, an excellent source of sulfur-containing amino acids. Finally, we have been instructed by doctors and other authoritarian sources to stay out of the sun and wear high SPF sunscreen whenever we do get sun exposure.

Another significant contributor is the high carbohydrate, low fat diet, which leads to excess glucose in the blood stream that glycates LDL particles and renders them ineffective in

delivering cholesterol to the tissues. One of those tissues is the skin, so skin becomes further depleted in cholesterol due to glycation damage to LDL.
9. Sulfur Deficiency and Muscle Wasting Diseases

In browsing the Web, I recently came upon a remarkable article [Dröge1997] which develops a persuasive theory that low blood serum levels of two sulfur-containing molecules are a characteristic feature of a number of diseases/conditions. All of these diseases are associated with muscle wasting, despite adequate nutrition. The authors have coined the term "low CG syndrome" to represent this observed profile., where "CG" stands for the amino acid "cysteine," and the tripeptide "glutathione," both of which contain a sulfhydryl radical "-S-H" that is essential to their function. Glutathione is synthesized from the amino acids cysteine, glutamate, and glycine, and glutamate deficiency figures into the disease process as well, as I will discuss later.

The list of diseases/conditions associated with low CG syndrome is surprising and very revealing: HIV infection, cancer, major injuries, sepsis (blood poisoning), Crohn's disease (irritable bowel syndrome), ulcerative colitis, chronic fatigue syndrome, and athletic over-training. The paper [Drage1997] is dense but beautifully written, and it includes informative diagrams that explain the intricate feedback mechanisms between the liver and the muscles that lead to muscle wasting.

This paper fills in some missing holes in my theory, but the authors never suggest that sulfur deficiency might actually be a precursor to the development of low CG syndrome. I think that, particularly with respect to Crohn's disease, chronic fatigue syndrome, and excessive exercise, sulfur deficiency may precede and provoke the muscle wasting phenomenon. The biochemistry involved is complicated, but I will try to explain it in as simple terms as possible.
I will use Crohn's disease as my primary focus for discussion: an inflammation of the intestines, associated with a wide range of symptoms, including reduced appetite, low-grade fever, bowel inflammation, diarrhea, skin rashes, mouth sores, and swollen gums. Several of these symptoms suggest problems with the interface between the body and the external world: i.e., a vulnerability to invasive pathogens. I mentioned before that cholesterol sulfate plays a crucial role in the barrier that keeps pathogens from penetrating the skin. It logically plays a similar role everywhere there is an opportunity for bacteria to invade, and certainly a prime opportunity is available at the endothelial barrier in the intestines. Thus, I hypothesize that the intestinal inflammation and low-grade fever are due to an overactive immune system, necessitated by the fact that pathogens have easier access when the endothelial cells are deficient in cholesterol sulfate. The skin rashes and mouth and gum problems are a manifestation of inflammation elsewhere in the barrier.

10. Summary

Although sulfur is an essential element in human biology, we hear surprisingly little about sulfur in discussions on health. Sulfur binds strongly with oxygen, and is able to stably carry a charge ranging from +6 to -2, and is therefore very versatile in supporting aerobic metabolism. There is strong evidence that sulfur deficiency plays a role in diseases ranging from Alzheimer's to cancer to heart disease. Particularly intriguing is the relationship between sulfur deficiency and muscle wasting, a signature of end-stage cancer, AIDS, Crohn's disease, and chronic fatigue syndrome.

The African rift zone, where humans are believed to have first made their appearance several million years ago, would have been rich with sulfur supplied by active volcanism. It is striking that people living today in places where sulfur is abundantly provided by recent volcanism enjoy a low risk for heart disease and obesity.

In my research on sulfur, I was drawn to two mysterious molecules: cholesterol sulfate and vitamin D3 sulfate. Researchers have not yet determined the role that cholesterol sulfate plays in the blood stream, despite the fact that it is ubiquitous there. Research experiments have clearly shown that cholesterol sulfate is protective against heart disease. I have developed a theory proposing that cholesterol sulfate is central to the formation of lipid rafts, which, in turn, are essential for aerobic glucose metabolism. I would predict that deficiencies in cholesterol sulfate lead to severe defects in muscle metabolism, and this includes the heart muscle. My theory would explain the protective role of cholesterol sulfate in heart disease and muscle wasting diseases.

I have also argued that cholesterol sulfate delivers oxygen to myoglobin in muscle cells, resulting in safe oxygen transport to the mitochondria. I argue a similar role for alpha-synuclein in the brain. There is a striking relationship between Alzheimer's and sulfur depletion in neurons in the brain. Sulfur plays a key role in protectiing proteins in neurons and muscle cells from oxidative damage, while maintaining adequate oxygen supply to the mitochondria.

When muscles become impaired in glucose metabolism due to reduced availability of cholesterol sulfate, proliferating fat cells become involved in converting glucose to fat. This provides an alternative fuel for the muscle cells, and replenishes the cholesterol supply by storing and refurbishing cholesterol extracted from defective LDL. Thin people with cholesterol and sulfur deficiency are vulnerable to a wide range of problems, such as Crohn's disease, chronic fatigue syndrome, and muscle wasting, because fat cells are not available to ameliorate the situation.

Cholesterol sulfate in the epithelium protects from invasion of pathogens through the skin, which greatly reduces the burden placed on the immune system. Perhaps the most intriguing possibility presented here is the idea that sulfur provides a way for the skin to become a solar-powered battery: to store the energy from sunlight as chemical energy in the sulfate molecule. This seems like a very sensible and practical scheme, and the biochemistry involved has been demonstrated to work in phototrophic sulfur-metabolizing bacteria found in sulfur hot springs.

The skin produces vitamin D3 sulfate upon exposure to sunlight, and the vitamin D3 found in breast milk is also sulfated. In light of these facts, it is quite surprising to me that so little research has been directed towards understanding what role sulfated vitamin D3 plays in the body. It is recently becoming apparent that vitamin D3 promotes a strong immune system and offers protection against cancer, yet how it achieves these benefits is not at all clear. I strongly suspect that it is vitamin D3 sulfate that carries out this aspect of vitamin D3's positive influence.

Modern lifestyle practices conspire to induce major deficiencies in cholesterol sulfate and vitamin D3 sulfate. We are encouraged to actively avoid sun exposure and to minimize dietary intake of cholesterol-containing foods. We are encouraged to consume a high-carbohydrate/low-fat diet which, as I have argued previously (Seneff2010), leads to impaired cholesterol uptake in cells. We are told nothing about sulfur, yet many factors, ranging from the Clean Air Act to intensive farming to water softeners, deplete the supply of sulfur in our food and water.

Fortunately, correcting these deficiencies at the individual level is easy and straightforward. If you just throw away the sunscreen and eat more eggs, those two steps alone may greatly increase your chances of living a long and healthy life.
References

1. Axelson1985
Magnus Axelson, "25-Hydroxyvitamin D3 3-sulphate is a major circulating form of vitamin D in man," FEBS Letters (1985), Volume 191, Issue 2, 28 October, Pages 171-175; doi:10.1016/0014-5793(85)80002-8

LCD Stops Bug Bites (5h Edition)

Even with all of the great benefits that I report here, we all hate mosquito bits and ticks are becoming a scourge with Lyme disease. Well guess what? Between the LCD and our water, which is impossible to separate rationally, This year, I get no insect bites and the ticks seem to drop off of me. I am just no longer good food for them. Admittedly, it has taken many years to arrive at this point, but I am liking it a lot. The mosquitoes still buzz me, mind you, but never a bite and I have been in the woods a lot and not a single tick has dug in so far.

Cheers1

57. Post Commentary:

Now that you have read this book, if you followed the links, you are more informed than most experts and you now can, "Take Charge of You Health," as Dr. Mercola says. Moreover, you can do this in ways that even the most informed health practitioners do not know about today. Mercola never put together the importance of HL MSM even though he has interviewed Dr. Steffanie Seneff on several occasions (?).

Note:
This book is for information only. It represents the results, observations, views, and opinions of author and those who have contributed and tried the various protocols, but is not a recommendation for treatment. Anyone reading it should consult his/her physician before considering treatment.

(5th edition)

Bibliography:

"An Introduction to Spontaneous Evolution" Bruce H. Lipton, PhD

"Back Hole" Nassim Haramein

"Beyond Pyramid Power," Dr. Patrick Flanagan, PhD
ISBN 0-87516-208-8

"Pyramid Power" Max Toth and Greg Nielson ISBN D-89281-106-4

"The Biology of Belief" Bruce H. Lipton, PhD
https://www.brucelipton.com/books/biology-of-belief

"White Paper" Dan Nelson, PhD Https://www.waybackwater.com/dans-white-paper

Books below by James Robert Clark addressing spirituality are available in Amazon books:

"Beyond Epigenetics"

- **Paperback 116 pages**
- **Platform 2nd Edition (July 13, 2015)**
- **ISBN-10: 1515060454**
- **ISBN-13: 978-1515060451**

"Methylation, Awareness and You"

- **Print Length: 114 pages**
- **Publication Date: December 19, 2014**
- **BASIN: B00R8P7R3**

"Pretty Fine Sex is Spiritual"

- **Paperback: 282 pages**
- **Platform: 1st edition (Sept 22, 2017)**
- **ISBN-10: 1978487894**
- **ISBN-13: 978-1978187870**

"Surrender to The Oneness"

- **Paperback: 68 pages**
- **Platform: (December 6, 2015)**
- **ISBN-10: 1519499345**
- **ISBN-13: 978-1519499349**

About the Author:

Jim holds a bachelor of Arts & Science from U of Delaware and a 1st Professional Degree in Architecture from the Boston Architectural College. His firm, James Robert Clark, Arch/ Planner, Inc is located in Milton, Delaware.

Jim is an accomplished artist working in many media. His paintings in oils, watercolors, and acrylics have been shown in local shows. His other hobbies include: tennis, skiiing, sailing, fishing, furniture design and woodworking, yacht design and he has built several boats. including a 26' diesel aux. sloop.

He is particularly interested in New Energy Technologies and has helped develop several new fuel saving devices including the CoilPack4U that is currently being marketed by IES Technogies, Inc.

CPSIA information can be obtained
at www.ICGtesting.com
Printed in the USA
LVHW091148240520
656464LV00002B/261